CAMBRIDGE SOCIAL BIOLOGY TOPICS

Series editors
S. Tyrell Smith and Alan Cornwell

The Food Resources of Man

David J.R. Millerchip
Senior Lecturer
Bulmershe College of Higher Education,
Reading, Berkshire

CAMBRIDGE
UNIVERSITY PRESS

CAMBRIDGE UNIVERSITY PRESS
Cambridge, New York, Melbourne, Madrid, Cape Town, Singapore,
São Paulo, Delhi, Dubai, Tokyo

Cambridge University Press
The Edinburgh Building, Cambridge CB2 8RU, UK

Published in the United States of America by Cambridge University Press, New York

www.cambridge.org
Information on this title: www.cambridge.org/9780521288910

First published 1984
Fifth printing 1992
Re-issued in this digitally printed version 2009

A catalogue record for this publication is available from the British Library

ISBN 978-0-521-28891-0 Paperback

Contents

Preface

Man, as an omnivore, is able to obtain nourishment from foods derived from a wide variety of plants and animals.

Until several thousand years ago, this food was obtained solely from wild plants and animals. Man did not cultivate plants, relying instead on those edible fruits, leaves, stems and roots that were available naturally. Similarly man caught, slaughtered and ate what wild animals he could, and did not herd or raise animals. This **hunter-gatherer** life-style was satisfactory – nutritionally – when the number of people was small. In these circumstances, the quantity of plants and animals consumed by man was negligible in relation to the total numbers and the reproductive abilities of the plants and animals remaining.

From the hunter-gatherer culture, there evolved gradually a more settled agriculture, which allowed for the social development of man and also provided an increased food supply. To obtain this increase in food supply, man altered the environments of the plants and animals that were useful for food production. For example, he saved seeds for later planting, cleared land to reduce competition from other, less desirable plants, and cultivated his plants in one place to allow easier tending and harvesting. Similarly, he herded together the more docile animals to provide a readily available meat supply.

By altering their environments man also altered the evolutionary selection pressures upon the organisms that he had chosen to domesticate. Thus they evolved – over the millennia – to become the plants and animals of agricultural importance that we know today. It is important to realise that they occupy niches very largely created and maintained by man. Without man's intervention, they would return to the wild. These populations would then probably have difficulty in surviving; those that did survive would represent a modern equivalent of their wild ancestors, and would be very poor sources of food for man. Thus the quantity and quality of virtually all the food produced today is dependent upon man's continuous intervention to maintain these altered environments.

This book looks individually at certain modern plants and animals of major agricultural importance and considers some of the systems by which they are husbanded.

Living in a country in which the amount of food available to each one of us rarely warrants a second thought, we tend to forget that most people in the world are hungry. The figure on page iv shows the average amounts of food energy and protein available to people in different countries. The vertical broken lines relate to the recommended minimum consumption per day. These can

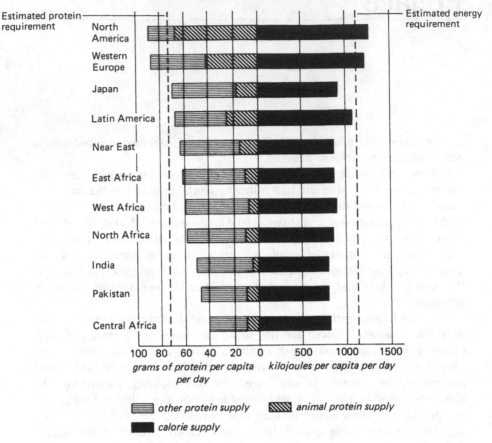

PROTEIN AND ENERGY INTAKE

Estimated protein requirement

Estimated energy requirement

North America
Western Europe
Japan
Latin America
Near East
East Africa
West Africa
North Africa
India
Pakistan
Central Africa

100 80 60 40 20 0   500 1000 1500

grams of protein per capita per day   *kilojoules per capita per day*

|||| *other protein supply*   \\\\\\ *animal protein supply*

■ *calorie supply*

only be informed guesses because food quality varies as does the metabolism of individual people. However, the histogram does emphasise the differences between **developed** and **developing countries**, and highlights the need for increased world food production.

This book therefore considers ways in which food production has been increased in recent years. However, it must be realised that purely scientific/agricultural methods of increasing food production do not necessarily mean an increase in the amount of food available to hungry people. Many detrimental economic, political and social factors contribute to the uneven distribution of food and to the poor nutritional status of people in developing countries, and these problems must be solved before hunger is eliminated.

1 Crops

1.1 Introduction

The preface indicated the differences in human energy and protein consumption in different parts of the world. It is important to realise that this energy and protein is supplied by different foods in these different countries. The developed countries enjoy a wide range of foods from a variety of plants and animals. However, this is not the case for many people in developing countries. They rely for a major part of their diet on one food. This food is referred to as a **staple food**. For example, in Central America the staple food is maize; in S. E. Asia it is rice, whilst roots and tubers (e.g. cassava, sweet potatoes) are staples for many in central Africa. There seem to be two reasons for this reliance on one staple food:

(a) the staple food is grown easily by most people in the areas concerned;

(b) the majority of people seem to be too poor to buy other foods.

Certainly Figure 1.1 indicates that diet changes with increase in income.

This figure indicates that generally the poorest people have the highest proportion of **carbohydrate** in their diet (proportion, not necessarily total quantity, because the total food available to them might be restricted). This high proportion of carbohydrate is because these staple foods consist mainly (excluding water) of carbohydrates. Other nutrients (proteins, vitamins, minerals) may not be present in adequate proportions. As income rises, the proportion of carbohydrates in the diet decreases because the higher income allows the purchase of foods derived from animals, which are always more expensive than those derived from plants. This is reflected in an increased proportion of animal fats and the change from vegetable to animal proteins.

Figure 1.1   ENERGY DERIVED FROM FATS, CARBOHYDRATES, PROTEINS
AS PERCENTAGE OF TOTAL CALORIES ACCORDING TO THE INCOME
OF THE COUNTRIES (1962)

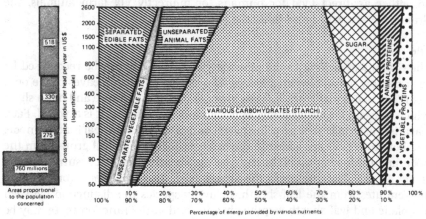

1.2 Staples

In terms of quantity produced and numbers of people fed, the most important staple foods in the world are **cereals**. This is because there are many species of cereals, each with a large number of varieties. This means that cereals are found worldwide in many different habitats. Also the flowering heads of the cereals produce a grain which will keep for a number of years with little deterioration when in the right storage conditions. The main cereals are **wheat, maize** and **rice** (approximately 25% each) followed by oats, sorghum and millets which together make up the remaining 25% (in decreasing order of approximate world grain production tonnage). In developing countries these grains will, in the main, be consumed directly by the people. Very little will be used to feed farm animals. The reverse is the case in developed countries. Very little is consumed by the people, but large quantities of cereal grains are used to provide a major part of the food of some farm animals (e.g. pigs, poultry). These farm animals then provide the protein and fats for the higher income peoples as shown in Figure 1.1.

Potatoes, sweet potatoes, yams, legumes (peas, beans, soya, peanuts etc.) and sugar cane are major **non-cereal** staples. Whilst each may be the staple for some peoples, they are not found worldwide because of their specific ecologies.

1.3 Gramineae

The majority of the cereals mentioned above belong to the plant family **Gramineae**. There are a few very minor exceptions, e.g. buckwheat, a member of the family Polygonaceae. The Gramineae also includes the **grasses**. The term grasses can be used to indicate that the vegetative parts of the plant are eaten by animals (i.e. are **forage** or **herbage**) and so does not include the cereals. Sometimes the word 'grasses' is used colloquially to mean **all** of the Gramineae. In this book the term Gramineae will be used when all members of the family are included and the term cereal will be used for those members of the family Gramineae that are grown for their grains. The term grasses will be used for those members grown primarily so that their vegetative parts can be eaten by animals.

Grasses have worldwide importance as the basic food for all grazing animals. So, either as direct food for humans, or indirectly via farm animals, the Gramineae are very important food suppliers.

The shoot

The Gramineae belong to the **monocotyledons** and so can be recognised by their narrow, parallel-veined leaves. The shoot consists of leaves borne on a stem. The bamboo (Figure 1.2) is a member of the Gramineae and has a shoot which is useful for initial study, although in some respects it is atypical. Each leaf arises at a **node**. The growing part (the **meristem**) of the leaf is at this node, and here new cells are added to the base of the leaf. The leaf grows from the node, and the lower part of the leaf (the **leaf-sheath**) completely surrounds the stem. As the leaf matures, the upper part, the leaf blade (or **lamina**), differentiates from the leaf-sheath. In many species the distinction between leaf-blade and leaf-sheath is quite obvious, and at the junction there may be

Figure 1.2

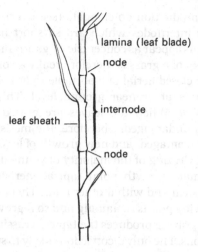

lamina (leaf blade)

node

internode

leaf sheath

node

two **auricles** and a **ligule** (see Figure 1.3). These are often distinctive features useful for recognition.

The bamboo has elongated **internodes** and it is this elongation that is not typical of the agriculturally important grazing grasses, though the flowering

BARLEY

lamina

ligule

auricle

leaf sheath

Figure 1.3

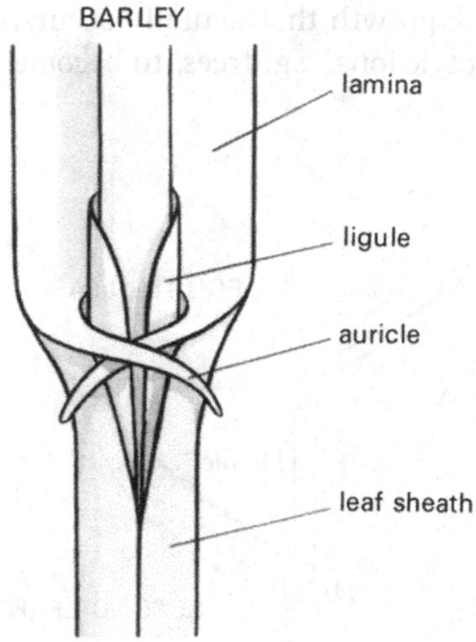

Common temperate cereals can be distinguished, in the vegetative stage, by the appearance of the ligule and auricles.

Barley has long auricles with no hairs (Barley: Big and Bald auricles). Leaf blades twist clockwise when observed from above.

Wheat has blunt auricles which are hairy or 'whiskery' (Wheat: With Whiskers). Leaf blades twist clockwise.

Oats has no auricles (Oats: O); anticlockwise twist to leaf blade.

Rye has short ligule and auricles (Rye: Restricted). Clockwise twist to leaf blade.

WHEAT

OATS

RYE

stems for seed/grain production do elongate (see later). In these grasses the vegetative shoot has internodes which are so short as to be nearly non-existent. The nodes thus appear as concentric rings around a minute stem. The visible part of the shoot of a grass is therefore really a collection of concentric leaves (much like the closed aerial of a portable radio). The leaves thus arise from a very short stem at or near ground level. This fact is of supreme agricultural importance. When the leaves are grazed, this short stem at ground level remains undamaged. Therefore, the **meristematic tissue** of the nodes also remains undamaged, and new growth of leaves (from the bottom) proceeds unhindered. Grazing of the majority of young dicotyledonous plants would remove the main growth point (**apical meristem**) from the stem, probably most of the stem and with it other buds. There would thus be no, or only one or two, growing points remaining and so regrowth would be slow or unlikely. This is why grazing produces the typical **grassland climax** with very few mature dicotyledons. The only dicotyledons likely to survive are those with either a **woody stem** (e.g. trees, bushes) or those with a **prostrate habit**, e.g. dandelion, plantain and daisy, where again the stem is very short and the leaves frequently remain procumbent. In some tropical areas, rainfall is limited and falls in only a short period of the year. Grass growth is therefore very variable – from virtually nothing for several months of the year, to a few weeks of luxuriant growth in the rainy season. It is therefore very difficult to match the feeding requirements of the animals to the growth that actually occurs. This would soon allow far less nutritious dicotyledons, e.g. trees, to become

Figure 1.4a

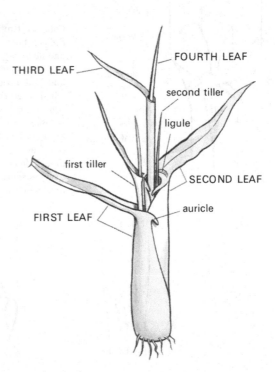

Sketch of simplified
cereal/erect grass to show
leaves and tillers.

established and severely reduce the grazing potential of the land. To control these nutritionally undesirable plants, the use of **fire** is an accepted management technique. Particularly in the savanna areas of Africa, controlled burning is used to remove old and dry vegetation of low feeding value. This burning also has the added advantage of stimulating the grass into a flush of growth during dry periods.

Tillers

In the same way that dicotyledonous plants have buds in the angle (axil) of the leaf, so **axillary buds** are present in the Gramineae. These buds grow to produce **axillary shoots** which have the same concentric leaf sheath structure as the main shoot. These axillary shoots are referred to as **tillers**. Grasses will often produce several tillers in rapid succession, which ensures quick vegetative growth, a feature that is particularly important in grasses.

Sexual reproduction in the Gramineae

After a period of vegetative growth, **flower formation** is initiated in the Gramineae. In temperate areas the initiation stimulus is **increasing day length**, though a prior period of cold is necessary for some species or varieties. This cold requirement (**vernalisation**) is particularly important in the cereals as it serves to differentiate the winter cereals from the spring cereals. Winter cereals must be sown in autumn prior to the cold temperatures of winter. Only after this cold

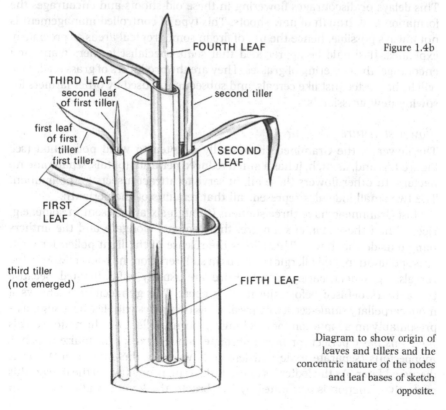

Figure 1.4b

Diagram to show origin of leaves and tillers and the concentric nature of the nodes and leaf bases of sketch opposite.

period will they then be able to respond to increasing day length, and thus flower. If sown in spring, they would receive no (or insufficient) cold and would not produce flowers – they would remain in the vegetative state. (Some winter cereals are really spring cereals which can survive the winter, but do not require cold to flower.) For spring cereals there is no such cold requirement. They are sown in spring and flower just in response to the increasing day-length stimulus.

When increasing day length initiates flower production, the production of new leaves ceases. Instead, an inflorescence is formed. The flowers on the inflorescence are borne above the vegetative shoots. This appears to facilitate **wind pollination** (see later).

Tropical Gramineae may be short-day plants or not sensitive to day-length changes since the length of the day alters very little near the equator.

Obviously flower initiation is important in the cereals – it produces the grain which is the prime reason for growing them. In grasses, as just indicated, no new leaves are formed on those shoots that are flowering and whilst those that remain may still grow, this could herald a decline in total vegetative growth as no new tillers are formed. This would mean a reduction in food for grazing animals. However, some vegetative growth will be taking place in those tillers which are too young (immature tillers) to flower. These will provide some grazing during the later months. If the numbers of animals grazing the grasses can be controlled, it is possible for many of the old (or first) shoots to be eaten. This delays or discourages flowering in these old shoots and encourages the formation and growth of new shoots. This type of controlled management is not always possible, hence the use of fire in some tropical areas as previously explained. It should be appreciated that some specialist farmers want, and encourage, the flowering of grasses. They are the producers of grass seed. This will be harvested just like cereals, and subsequently used by other farmers for sowing new grasslands.

Flower structure

The flower of the Gramineae is basically adapted to wind pollination (see Figure 1.5) and, as such, it has a much reduced **perianth** (petals/sepals) and no nectary. In other flowers these might serve to attract insects for pollination. The two small **lodicules** represent all that remains of the perianth.

Most Gramineae have three **stamens** (though six are present in some, e.g. rice). When these stamens mature, the filaments elongate and the **anthers** hang outside the flower. This allows the release of the light pollen for wind cross-pollination, and allergic reactions in sufferers from hay-fever. Most of the cereals, e.g. wheat, barley, oats and rice, are usually self-pollinated, because the anthers **dehisce** before the flowers open. The **gynaecium** is always a monocarpellary single-seeded ovary and has two styles with feathery stigmas – presumably an adaptation for catching air-borne pollen. All the major cereals have this basic perfect (or **hermaphrodite**) flower, except for maize which is **monoecious** (separate male and female flowers on the same plant) and is discussed later. As the lodicules afford little or no protection to the flower, this protective function is performed by two **bracts**, the **lemma** and the **palea**. In

Figure 1.5

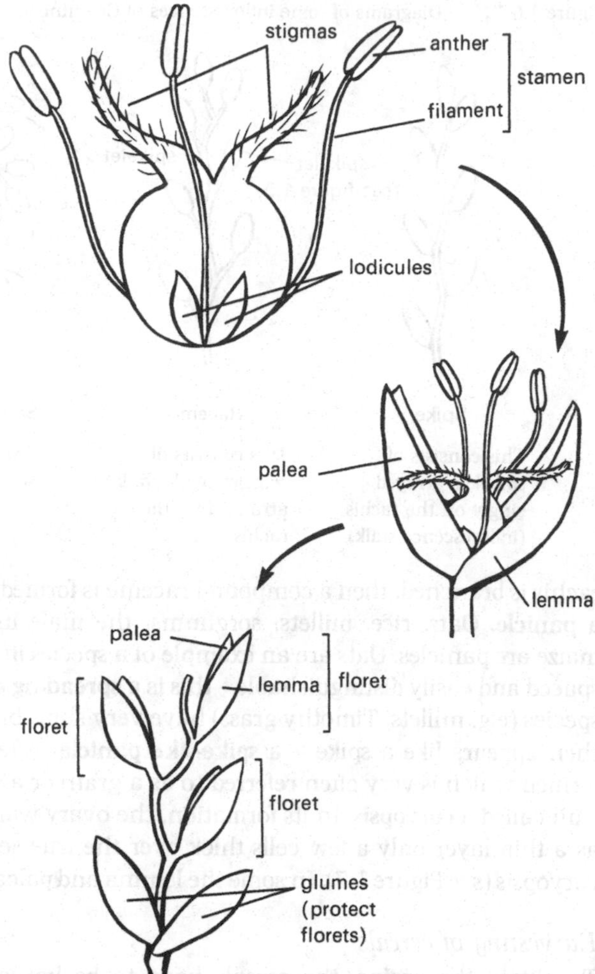

stigmas

anther ⎤
      ⎬ stamen
filament ⎦

lodicules

Grass flower

palea

Grass floret, open.
A floret consists of a flower
together with the protective
lemma and palea.

lemma

palea

lemma ⎤ floret

floret

floret

glumes
(protect
florets)

Several florets together
form a spikelet:

some plants, the lemma may bear a barbed projection called an **awn**. The flower, plus the lemma and the palea, are together called a **floret**. Several florets may be arranged together into a **spikelet** which will have protective **glumes** (see Figure 1.5). The number of florets per spikelet and the distribution of spikelets on the flower head (**inflorescence**) vary from species to species. (See Figure 1.6.)

The inflorescence

The different inflorescences produced by the Gramineae can be regarded as variations upon a theme (see Figure 1.6). If the spikelets are attached directly to the **rachis** (central stem) with no stalk (**sessile**) the inflorescence is called a **spike**. Wheat, barley, rye and the 'cob' of maize are typical spikes. If the spikelets are attached by a stalk to the rachis, then the inflorescence is a **raceme** (not common in Gramineae). If, however, each of the stalks attached to the

Figure 1.6   Diagrams of some inflorescences of Gramineae

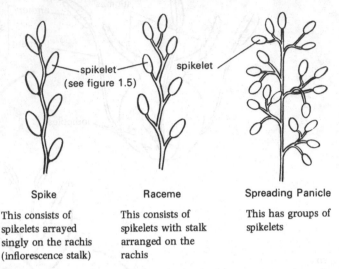

Spike	Raceme	Spreading Panicle
This consists of spikelets arrayed singly on the rachis (inflorescence stalk)	This consists of spikelets with stalk arranged on the rachis	This has groups of spikelets

rachis is branched, then a compound raceme is formed which is usually called a **panicle**. Oats, rice, millets, sorghums, the male inflorescence (tassels) of maize are panicles. Oats are an example of a species in which the spikelets are spaced and easily distinguishable – this is a **spreading panicle**. However, some species (e.g. millets, Timothy grass) have very short branches and the panicle then appears like a spike – a **spike-like panicle**. After fertilisation a fruit is formed which is very often referred to as a **grain** or a **seed**. Botanically, it is a fruit called a **caryopsis**. In its formation, the ovary wall (the **pericarp**) adheres as a thin layer only a few cells thick over the true seed, and thus forms the caryopsis (see Figure 1.7). In some the lemma and palea also remain (see later).

Harvesting of cereals

To obtain the grains, the cereals have to be harvested. With little or no mechanisation, this is basically a two-stage process. First, the mature plant stalks are cut, usually low down near to the soil. These are then gathered together (often into bundles, e.g. sheaves) and, after possibly a further period of maturation in the field in groups (sometimes called stooks or shocks), these are collected together in a stack or barn.

The second process (which may take place very soon after the first, or be delayed some time) involves the separation of the grain from the remaining plant material. This involves **threshing** and **winnowing**. Beating with sticks, trampling by cattle or man, and flailing are primitive threshing methods still used worldwide. Throwing the resultant products up in the air, and allowing the wind to remove the lighter parts, is the most primitive form of winnowing. Mechanisation initially brought the **threshing machine** (which also winnows). This separates the grain by the rotation of a cylinder of grooved bars closely adjacent to a static concave reticulate framework. Sieving, and then winnowing (by mechanical fan), separates the grain from the lighter parts (the **chaff**, composed of loose glumes, lemmas and paleas) and the **straw** (flower stalk). The most technically advanced farmers are able to combine these two

Figure 1.7   Diagram of wheat caryopsis (vertical section)

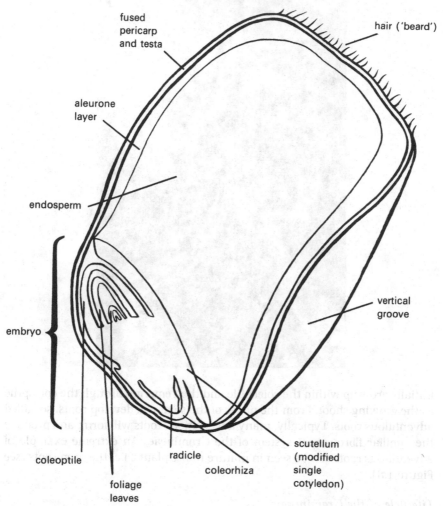

processes (cutting and threshing/sieving/winnowing) in one mobile machine, the **combine harvester**.

The grains produced by these post-harvest operations can be classified into two groups – those with **naked caryopses**, and those with **covered caryopses**. Wheat, rye and maize are examples of naked caryopses, and barley, oats and rice are examples of covered caryopses. The covered caryopses still have the lemma and palea surrounding them which are referred to as the **hull** or **husk**. Before human consumption, rice is dehulled (see later).

Germination and root production

The wheat embryo (typical of many Gramineae) shows the primary root (the **radicle**) enclosed by the **coleorhiza** (see Figure 1.9). Upon germination, the coleorhiza increases only slightly in length before the **primary root** pushes through, followed by the **lateral roots**. Meanwhile the **coleoptile** will have elongated and pushed above the surface of the soil. The first foliage leaves will

Figure 1.8

initially grow up within the coleoptile and then emerge through the coleoptile as the growing shoot. From the nodes of the stem will develop roots, so called **adventitious roots**. Typically, many adventitious roots will form, and produce the familiar **fibrous root** system of the Gramineae. An extreme example of adventitious roots can be seen in mature maize plants, i.e. the prop roots (see Figure 1.8).

Lifecycle of the Gramineae

The cereals are all grown as annuals. (There is some evidence that some plants could persist from year to year – but only in totally uneconomic quantities.) Figure 1.9 shows the life-cycle of wheat.

The grazing grasses of agricultural importance could probably be grown as annuals. If not grazed, they would flower and produce seed within a year. However, they nearly all persist from year to year by vegetative means and therefore could be classed as **perennials**. Perennation in these grasses mainly depends upon the production of tillers which form their own adventitious roots, and the subsequent survival of these tillers over winter. In some varieties of Italian ryegrass, one of the most important temperate cultivated grasses, this ability to survive is poorly developed, and thus these pastures are rarely very productive after the first year. However, foliar production in the first year is excellent and highly nutritious and so the grass has attained importance as fodder for high yielding cows. Some grasses which are only important in agriculture because of their nuisance value (e.g. the weed *Agropyron repens* commonly known as wicks, twitch, couch) perennate by **rhizomes** (which are basically underground stems) or **stolons** (similar but on the surface). Rhizomes and stolons seem to originate as horizontal tillers. They grow out sideways, breaking through the parental leaf-sheaths, into/onto the soil, and their stems subsequently elongate to perennate the plant. At the nodes of these horizontal stems, adventitious roots are formed and then new leaves will arise to form new plants. No rhizomatous or stoloniferous grass has high agricultural productivity. Compared with pasture grasses, it seems that most of the photosynthetic products go into the perennating rhizomes and stolons, and little into the foliage.

Figure 1.9

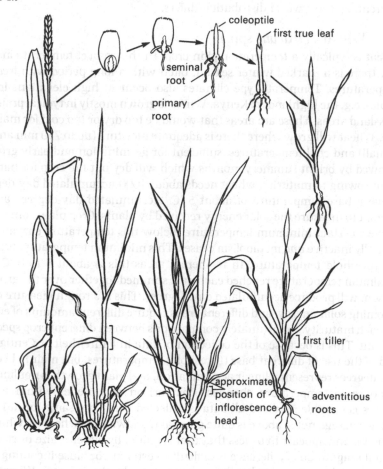

Life cycle of wheat.
The germinating grain produces a coleorhiza through which the primary root soon develops. The coleoptile appears above soil level, and then the first true leaf breaks through the coleoptile. Tillers develop within the leaf sheaths (refer also to figure 1.4). Eventually flowering is initiated, and the inflorescence develops within the centre of the leaves, finally to appear and produce more grain.

2 Cereals and other crops

The worldwide distribution of the cereals, as with other crops, is directly related to their physiological reaction with their environment. This might be termed the ecology of the crop. As the ecologies of the main cereals are different, so their world distribution differs.

2.1 Wheat (*Triticum* spp.)

Wheat is typically a **temperate region** crop. In this respect temperate implies that there is a marked winter season, often with a long period with freezing temperatures. Temperate-type climates also occur at high elevations in the tropics, e.g. the highlands of Kenya. Wheat is grown mostly in typical **prairie** or **grassland** areas. These are areas that would be too dry or too cold for maize or rice. Wheat will grow where there is adequate moisture (up to 750 mm annual rainfall) and cool temperatures, sufficient for germination and early growth, followed by bright summery months which will dry out the crop for harvest. From sowing to maturity, wheat needs about 1000 accumulated day-degrees above a base temperature of about 5 °C. Accumulated day-degrees are an attempt to measure the solar energy received by plants. Crop plants only grow above a certain minimum temperature. Below this temperature enzymes are virtually inactive and metabolism ceases. This minimum temperature is called the threshold temperature. In temperate areas this is about 5 to 6 °C. The maximum temperature reached each day, summed together over the growing season, will produce accumulated day-degrees. This is a rough measure of the incoming solar energy, and different crops require different amounts of energy to reach maturity. Unfortunately comparisons between different crop species is difficult. This is because of the collection of data in Fahrenheit or Centigrade, and of the use of different base (threshold) temperatures, but mainly because day-degrees represent a linear accumulation, and plant growth is not linear in response to temperature changes. Comparison of varieties within a crop species is possible provided all data is collected using the same formula.

The average noted above is quite general worldwide for wheat (of whatever variety), and appears to be less that that required by either maize or rice (the actual magnitude of difference is difficult to estimate because incoming solar energy has been measured in different ways for the three crops). Wheat does not grow well in the more tropical areas as high temperatures are unsuitable for its physiology as it is a C_3 plant. In C_3 plants both photosynthesis and photorespiration are stimulated by rise in temperature. At high temperatures, on still, hot days, photorespiration may exceed photosynthesis, and so the crop may decrease (Moore, 1981).

Though quite hardy to winters, especially under a blanket of snow, wheat plants will not survive the more severe winters of the higher latitudes (e.g. Canadian Prairies), and so only spring varieties can be sown in these regions. In lower latitudes (e.g. southern wheat belt of the USA) winter varieties, that is those which are autumn sown, can be grown. Under similar conditions, winter varieties generally yield more than spring varieties. Whilst this may be a genetic difference, it is probably due simply to the fact that they are growing for a longer period than the spring-sown varieties. In the UK, wheat is mainly grown in southerly parts, for the northerly regions tend to have too much moisture and insufficient sunshine hours for successful wheat growth. Traditionally, oats have been grown in these northerly areas, for they give an adequate yield in poor climatic conditions. The decline in the number of horses, coupled with increased yield of both barley (used for malting, brewing and stock-feed) and wheat has led to an increased acreage of barley and wheat in many parts of the UK at the expense of oats. Wheat has a major use as direct feed for humans. It is the only cereal that can successfully be used to make leavened bread. Whilst other cereals contain proteins, only certain varieties of the wheats have sufficient of the specific complex of proteins usually referred to as gluten to make them useful for making bread. Gluten gives a sticky dough that traps the carbon dioxide produced from yeast (or baking powder). These 'breadmaking' wheats are usually referred to as hard or strong wheats. Hard means that the **endosperm** (see Figure 1.7) yields a free-flowing, coarse and gritty flour during milling. Strong refers to the ability to produce good bread and is usually associated with a high protein content, in the order of 13–16%. Hardness and strength are different genetically controlled characters, yet almost invariably a hard wheat is also a strong wheat. These hard wheats usually come from areas of limited rainfall and hot summers (e.g. northern America; USSR). Soft wheats (i.e. fine flour, but not free-flowing when milled) are usually weak and lower in protein content, about 8–11%. These weak, soft wheats on their own do not produce a dough which will satisfactorily trap carbon dioxide and are therefore unsuitable for breadmaking. Whilst some will be used in the grist (mix of flour etc.) for breadmaking, they are more likely to be used for processing into animal feedstuffs, though a considerable quantity is used in the UK for biscuit making and confectionary manufacture. Soft wheats come from areas with higher precipitation and humidity, e.g. southern USA and Europe including the UK. The UK has to import all its strong wheat for breadmaking, traditionally from the Canadian Prairies. The UK grist has usually had to contain a greater proportion of imported strong wheat than native soft wheat. However, recent improvements in breadmaking processes have meant that a lower overall protein level can be used, with a consequent saving on expensive imported strong wheat. The major breakthrough in this breadmaking technology came with the Chorleywood Bread Process in 1961. This substituted the lengthy fermentation (still typical of home breadmaking) with a short vigorous mechanical mixing, which produced the necessary elastic properties of the proteins to allow successful trapping of carbon dioxide. Chemicals are also used to assist in this rapid production of 'elastic proteins'.

If wheat were being consumed as a staple food, the quantity of protein (8–16%) would provide a substantial proportion of the human protein requirement (for amount of protein required, see discussion later under 'other

Table 2.1 The essential amino acid content of cereal and legume seeds, and animal sources compared with WHO recommended levels

Amino acid	WHO recommendation (g/100 g protein)	Animal			Legume			Cereal					
		Hen's egg	Cow's milk	Beef muscle	Mung bean	Broad bean	Soya bean	Wheat	Rye	Barley	Oats	Maize	Rice
Cysteine and methionine	3.5	5.5	3.4	3.8	1.2	1.5	1.6	3.9	4.6	4.6	6.5	5.1	4.4
Lysine	5.5	6.4	7.9	8.7	8.1	6.5	6.4	2.0	3.3	3.1	4.4	3.5	4.0
Isoleucine	4.0	6.6	6.5	5.3	3.6	4.0	4.5	3.6	3.6	3.6	3.8	3.6	4.7
Leucine	7.0	8.8	10.0	8.2	7.0	7.1	7.8	6.7	6.7	7.2	7.7	11.6	8.5
Phenylalanine and tyrosine	6.0	10.0	10.2	7.5	7.4	7.5	8.0	7.7	7.0	8.2	7.8	7.2	10.3
Threonine	4.0	5.1	4.7	4.3	3.3	3.4	3.9	2.7	3.4	3.3	3.7	3.9	3.8
Tryptophan	1.0	1.6	1.4	1.2	n.a.	0.9	1.3	1.1	1.8	2.0	2.0	0.9	1.2
Valine	5.0	7.3	7.0	5.5	4.1	4.4	4.8	3.7	4.4	4.6	5.0	4.9	7.0

n.a. = not available.

crops'). However, the quality of the protein – as indicated by the proportions of amino acids – is also important. Whilst different wheats vary in the amino-acid composition of their proteins, figures generally indicate that they contain too little lysine to be an ideal protein source for humans.

Table 2.1 shows that in a comparison between proteins in wheat and the proteins of a more 'complete' food, such as egg, there is far less lysine present (in proportion) compared to the other amino acids. Egg protein is the usual standard with which to compare other protein because it seems to have nearly ideal proportions of amino acids for humans and is readily available and metabolised (compare with World Health Organisation recommendations). Wheat protein contains only about one third of the lysine content of egg protein. In areas where wheat is the staple food, there may be insufficient intake of lysine. Consequently, there will be retardation of protein metabolism, manifesting itself as the symptoms of malnutrition.

The modern milling process produces:

white flour – from the endosperm;

bran – from the pericarp, testa, and aleurone layer of the endosperm;

germ – embryonic tissues.

Bran is particularly rich in fibrous material. This fibrous material is little affected by intestinal digestive processes, but is particularly effective at absorbing water. Addition of bran to a human's diet increases the bulk and consequently the speed of transit of undigested material. It is speculated that this more rapid passage of faecal material reduces the incidence of certain intestinal diseases. This is one of the bases of the call by some for a return to such unrefined foods as wholemeal bread. This bread is made from **wholemeal flour,** i.e. flour from which nothing has been removed. This milling is traditionally done between two mill stones which grind the grains and also mix the bran with the white endosperm to produce brown flour.

The wheats referred to so far have all been varieties of *Triticum aestivum* (also called *T. vulgare* and *T. sativum*). There are other species of wheat which are grown on a much smaller scale, e.g. *T. durum* for flour to produce pastas such as macaroni, spaghetti and noodles.

2.2 Rye (*Secale cereale*)

The only other cereal with any breadmaking possibilities is rye, though rye bread is usually made from a mixed rye/wheat grist. In developed countries, rye finds its main use in the manufacture of various crispbreads. The rye plant has the advantage of greater winter hardiness than wheat, and it therefore can be grown further north than any other cereal. It is able to give a reasonable yield on poor soils with the minimum of attention and has, particularly in the past, been the staple of a few of the less developed communities of Europe and Asia.

It has been possible to produce a hybrid between wheat (*Triticum*) and rye (*Secale*) called *Triticale*. The varieties of *Triticale* now being grown have 42 chromosomes and are derived from durum wheat (28 chromosomes) and rye (14 chromosomes). They thus have a genetical resemblance to bread wheat (*T. aestivum*) which also has 42 chromosomes.

One of the aims of the production of *Triticale* was to obtain a cereal with the advantages of both parents. This has been partially achieved. The varieties of *Triticale* generally available have the hardiness and higher lysine content of rye (compare wheat and rye in Table 2.1). Whilst they produce a higher yield than rye, it is not – on average – quite as high as wheat. Several hundreds of thousands of hectares are grown throughout the world, with the vast majority being used as stockfeed. It seems as though it will be some time before it gains an increasing use as a staple for man, where the higher lysine content would be beneficial.

2.3 Barley (*Hordeum* spp.)

Whilst barley is a major cereal in the UK, it has less worldwide importance. Its ecology is similar to that of wheat, needing a temperate climate with adequate but not excessive rain and warm summers, but it is only consumed directly by man in small quantities as it lacks the ability to produce leavened loaves. Its major uses are for malting/brewing and for stockfeed.

It is interesting that the climate which produces good bread does not necessarily produce a barley that is satisfactory for brewing and malting. Whilst there are a few exceptions, in general the prairie climate that produces the higher protein wheats, also produces higher protein varieties of barley. However, this higher protein is not always desirable in barley for brewing and malting. First, the greater quantity of protein that is present is more likely to produce a cloudy, rather than a clear product (e.g. whisky, beer etc.). Secondly, the higher protein content usually means a lower carbohydrate content. With less carbohydrate there is less substrate for the production of alcohol, so there is less yield of end-product. Barley grown in the UK generally has a lower protein content and consequently a higher carbohydrate content. However, the use of nitrogenous fertilizers to produce higher yields of grain will often result in an increased protein content. This may lead to rejection by the maltsters, and the grain can then only be sold for animal feedstuffs, often at a lower price.

The agricultural policies of successive UK governments after the Second World War produced conditions in which the growth of cereals such as barley could be profitable. The resulting increase in the availability of barley coincided with developments in the knowledge of the feeding of cattle. Traditionally offspring of dairy cattle (e.g. Friesian, the main dairy breed of UK) had not been regarded as beef animals (providers of quality meat). However, particularly if crossed with traditional beef breeds, e.g. Hereford, they proved to have fast weight gain and produce high quality meat if given the appropriate food at the correct time. It thus became profitable to produce beef fed mainly on barley. Whilst previously barley may have constituted a small part of the diet of cattle it had never been a major part. The practice of feeding large quantities of barley to beef animals has now diminished, but the practice highlighted the importance of matching food quality and quantity to the inherent potential of the animals.

2.4 Oats (*Avena* spp.)

Usually a crop of the moist temperate lands, oats will give some yield under conditions of poor soil, climate and management. Improved conditions produce only a small increase in yield, and so other cereals capable of higher yields (e.g. wheat, barley and maybe maize) will be grown when these better conditions prevail. Throughout the world, most is used locally as stockfeed, though a small proportion is eaten directly by humans, for example as porridge, and consequently little enters world trade.

Oats are easily recognisable, when mature, by the spreading panicle (see Figure 1.6). However, do not assume that all you so identify in the UK are desirable. The troublesome weed 'wild oats' (*Avena fatua*) can be seen in the late spring/early summer with its inflorescences above the other cereals, e.g. in wheat and in barley crops. It obviously grows well in conditions very similar to other cereals, and is troublesome because it both contaminates the harvested crop, and tends to shed its ripe grain before the major cereal is harvested. These grains can remain dormant for years, to grow again later to contaminate another cereal crop. Herbicides to combat wild oats have been difficult to develop because of the similar physiology and ecology of wild oats to the common cereals. It was less of a problem under less intense cereal cropping and with crop rotations which allowed hand and mechanical weeding (see later p. 27 *et seq.*).

2.5 Maize (*Zea mays*)

Maize, with its possible origins in the Americas, is often referred to, worldwide, by its American synonym **corn**. As such it is the derivation of, for example, sweet corn, pop corn, cornflour, cornflakes, corn syrup – all maize products. In the UK, however, the word corn refers to all the traditionally grown cereals, wheat, barley, oats and rye, so that a UK cornfield will be growing any one of these cereals, whilst an American cornfield will be growing only maize. The word grains may be used in America to refer to cereals other than maize.

Possibly indicative of its American origin, about half the world production of maize grains comes still from the USA, the other half from the rest of the world where conditions are suitable.

Maize is intolerant of frost and so this is one limit to its worldwide distribution. Traditional varieties have a minimum temperature requirement for both germination and growth that is somewhat higher than the temperate cereals previously discussed. This requirement is about 9–10 °C, whereas wheat and barley require a minimum of about 4–5 °C, with rye possibly a little lower at about 2–3 °C.

Maize will grow well under moist conditions, though fungal attack may then be more prevalent. It is fairly tolerant of drier conditions as it is able to put down roots to over one metre. It will respond well to long periods of sunshine and high temperatures, producing correspondingly high yields. However, pollination can be poor if too little moisture is available at this crucial time. The germinated pollen grains have to produce pollen tubes of many centimetres length to effect fertilisation. Low pollination rates will mean low yields.

These requirements of maize have limited its spread into temperate areas. The relatively short growing season in the UK (between spring and autumn frosts) and only comparatively moderate summer temperatures mean that the production of mature hard grains suitable for storage and subsequent milling is not possible in the UK. These lower temperatures allow the plants usually to produce only soft immature grains. However, improved varieties have allowed an increasing area in the UK of maize for forage, as well as for human consumption – sweet corn and corn on the cob. The forage maize is usually mechanically harvested, chopped and stored as silage for feeding to cattle during the winter. The yield can be higher than from conventional grass-made silage. Also, it will be made in autumn (i.e. only when the plant has achieved maximum yield), often a less busy time of the year.

Morphologically, maize is unique among the cereals, and so it is usually easily recognised after only a brief acquaintance.

Its broad parallel-veined leaves are easily distinguished in young plants from the other cereals which have narrower leaves. The single main stem of the adult terminates in an inflorescence (of the type known as a panicle) bearing only male (staminate) flowers, and is referred to as the tassel. The male flowers are similar to the typical Graminaceous flower shown previously, except that they lack the female parts.

The female inflorescences are referred to as **cobs** and are borne in the axils of some of the middle leaves. The axis of the cob bears many rows of female (pistillate) flowers which are similar in appearance to the typical Graminaceous flower, except that the male parts are missing, and the ovary carries a style which elongates considerably during growth. The many styles emerging from the enclosing leaves of the cob are referred to as **silks** and some will achieve lengths of two, three or more decimetres before pollination. After pollination and fertilisation, the silks wither and the ovaries enlarge to form the typical maize grain (botanically a fruit called a caryopsis).

When grown to maturity in hot climates, the leaves enclosing the cob will dry, yet still remain attached. Similarly, the hard, mature grains remain attached to the axis of the cob. The chances therefore of natural 'seed' dispersal are negligible. Maize is therefore regarded as one of the prime examples of plants which have become so different from natural ancestors as to be unable to survive without considerable effort on the part of man.

However, one advantage of **monoecy** is to allow the comparatively easy removal of the tassels of selected plants. This facilitates cross-breeding between different maize strains or varieties, and has allowed selective breeding techniques to produce increases in yield, particularly in the USA, of several hundred per cent. This ease of emasculation is not possible in the other major cereals. It is a much more difficult task, involving patience, good eyesight, fine forceps and considerable skill to be able to remove anthers and yet leave the flowers still capable of being pollinated.

The breeding technique depends upon the production of **inbred lines** (plants with much more similar genotypes than the average plant) which are then crossed to exploit **hybrid vigour**. These plants then yield better than either of the parent lines. The only disadvantage to the farmer is that, like all 'F_1 hybrids', they fail to breed true in the next generation. If used as seed, they would produce a wide variety of plants with an overall much lower yield. So the farmer must continually buy new 'hybrid' seed – as must all the gardeners in this country who want the advantage of F_1 hybrid vegetables and flowers.

Most of the maize in the USA and other developed countries is used as

Figure 2.1

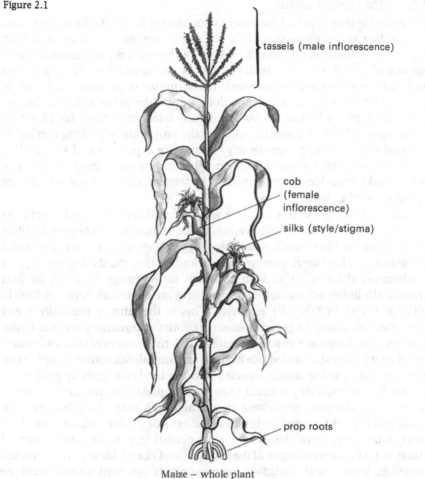

tassels (male inflorescence)

cob
(female
inflorescence)

silks (style/stigma)

prop roots

Maize – whole plant

stockfeed – both cobs and whole green plants. It is possible that 'corned beef' comes from cattle fed on maize (corn). It might, but the *Oxford English Dictionary* suggests a much earlier use of the word 'corned' before the American colonisation when it meant slightly salted meat. Also 'corned' can mean 'horned'. A little maize is used in developed countries for human consumption, with, e.g., specific varieties for sweet corn (richer in sugar), and pop corn (bursts on heating). Other corns are manufactured to yield starch (which may be turned into corn syrup or adhesives), the embryo may be crushed to yield corn oil for cooking, whilst some maize is used to prepare alcoholic drinks.

In the developing countries the proportion eaten by humans is much greater and in some places it is the staple food. It is then consumed in various ways; as flour for subsequent baking, or soaked and then cooked. Nutritionally, maize is similar to the other cereals, with a mainly carbohydrate grain and about 10% protein. However, this protein is normally low in both lysine and tryptophan (see Table 2.1). Plant breeders are attempting to produce maize with higher amounts of these amino acids.

2.6 Rice (*Oryza sativa*)

It is probable that over half the world's population is wholly dependent upon rice as their major foodsource. It is the staple in some of the areas of high population density, e.g. China, S. E. Asia. Of the three major cereals (wheat, maize and rice) it is the one which is mainly consumed directly by man and with little post-harvest preparation. (Most maize is processed and fed to animals; wheat is milled, made into dough and then baked before consumption.) Most rice will just be dehulled (see later) and then boiled before consumption. For many people, boiling is the only feasible cooking method as fuel and major cooking utensils (e.g. ovens) are expensive and scarce. This direct consumption by man also means that only a small proportion of rice enters world trade. Most is consumed by the farmers, or within a short distance of the place of growth.

Rice is a plant of the tropics. Here will be found the consistent short days which most varieties need for flowering, though some seem to be non-sensitive to day-length. Rice needs relatively high temperatures and prolonged consistent sunshine for maximum growth and rapid maturity. Figures suggest a minimum of about 20 °C, with optimum in the range 25–35 °C, for best growth. Minimum accumulated temperatures are quoted as 'between 3000 to 4000 °F' (1130–1500 °C) for maturity. This is the sum of the daily mean temperatures during the growing season. It is almost legendary that rice is also semi-aquatic, for we are all surely familiar with the pictures of men and water-buffaloes working the flooded rice fields. Monsoon rainfall usually supplies this water, which is subsequently directed into the level rice fields by man-made canals. Rice is frequently referred to as padi (or paddy) in the Far East, hence paddy fields. Monsoon conditions are usually such that rain falls copiously, stops, and is then followed by bright cloudless skies. Areas with similar total precipitation may have clouds for longer periods (e.g. some coastal parts of Malaysia). This cuts out some of the sunlight and rice yields are lower in what are otherwise excellent conditions. For most plants, constant water around the roots would produce typical waterlogged symptoms and eventual death due to inadequate gaseous exchange between the roots and the environment. Rice plants (like other **hydrophytes**) have tissues with intercellular spaces in the stems and roots (**aerenchyma**). It is believed that these tissues allow diffusion of air between the atmosphere and the root cells via leaves and stems, and thus help to overcome the poor gaseous exchange between roots and standing water. Some have suggested, with little apparent evidence, that maybe the respiratory physiology of rice roots is somehow unique and that the plant grows well on very reduced quantities of oxygen and under an accumulation of carbon dioxide greater than normally tolerated by plant roots.

There are many hundreds of varieties of rice throughout the world. Most of these are subspecies of *Oryza sativa*. However, there is evidence of rice from other species of *Oryza* being grown, some even being traded, in Africa and South America. It is possible that small communities may still be subsisting on **wild rice** which, it is speculated, will be very similar to plants chosen for domestication thousands of years ago. This plethora of varieties also means that different habitats will have been, and will still be, exploited, and therefore growing conditions between varieties will be different. About 10% of the rice that is grown is usually described as **upland rice**. It is usually grown at higher altitudes in a similar manner to the other cereals. It requires very little more water than normal cereals and is not grown in flooded fields. The yield of these varieties is said to be less than the **lowland**, flooded-field varieties.

Cultivation of the lowland rice usually involves initially sowing the seeds in seed beds and later transplanting the young plants into the flooded fields. This is a traditional labour-intensive method involving very little mechanisation. Recently, small tractors have begun to be used for the cultivation of the paddy fields, in place of the buffaloes and other draught animals.

Rice seems to yield well on all types of soils, and the main criterion for the field is that it will retain the flood water. Obviously clayey/silty soils are more likely to retain water, but the usual practice of working soil whilst it is moist or flooded produces a well-worked (puddled) soil which tends to lose its structure, and therefore pan, which restricts percolation.

Some rice is also grown in Europe. For example, rice is grown in the Camargue delta of the Rhone on land which was formerly marsh. After the Second World War, rice was grown mainly to desalinise ground prior to the planting of vines. However, rice yields were promising, and the commercial success of rice growing has been encouraged by irrigation schemes, as well as by the introduction of new varieties of rice. The town of Arles has thus become the centre of a rice-milling industry. For a long time France was self-sufficient in rice from the Camargue, though there has been a slight regression accompanied by a fall in labour demands as the growing of the crop has become more mechanised.

In the USA, where about 1% of the rice is produced, cultivation is highly mechanised. Often seeds are sown into unflooded fields using the typical tractor/seed drill machinery. The fields are then flooded, later to be drained prior to mechanised harvesting by combine harvester. However, some American farmers have even found benefit from sowing seeds by air into the already-flooded fields, though the wind and the low density of the seeds are problems.

Harvestable rice is produced in about four to five months, so more than one crop per year is feasible in the tropics with suitable varieties. The seed produced is the usual Graminaceous fruit called a caryopsis. In rice it is of the covered type, the adherent lemma and palea often being referred to as the **hull** (or husk). The hulls are removed before eating. Various primitive methods involve some form of pounding – treading by man or livestock, or water-powered devices to pound the rice against the ground. This produces not only separated hulls and kernels (caryopses), but much broken grain. Separation of hulls from kernels is then by winnowing. The whole kernels with the hulls removed are known as **brown rice**. Brown rice is consumed as such by most people. However, in the developed countries, most people prefer (or are only able to obtain) white rice. This has been further mechanically treated to remove the pericarp, testa and aleurone layer of the endosperm, and then further treated to shine or polish the endosperm.

The nutritional value of brown rice is far greater than white, as the removed parts are comparatively high in protein, vitamins and minerals. A process which is sometimes used to assist dehulling also has nutritional advantages should subsequent conversion to white rice also take place. This process is parboiling. The rice is placed in boiling water for just a few minutes. This is insufficient to cook the rice for human consumption but allows a much easier separation of the hull from the kernel. Its nutritional advantage is that the small amount of inward movement of water drives most of the vitamins from the outer layers into the endosperm. Then the high temperature of the water gelatinises some of the starch of the endosperm to seal partially the rice kernel, thus offering a longer storage life. If subsequently made into white rice, the nutritional content is then higher than without the treatment. Those interested should read about the disease beri-beri, the incidence of which is related to the eating of white (or polished) rice.

2.7 Other crops

The staple crops previously discussed are valuable mainly for the carbohydrate which they contain. There is usually insufficient **protein** to meet all human requirements. Also, as indicated previously, the protein is usually deficient in, at the least, lysine.

The estimated human requirement for protein has dropped quite considerably with advances in knowledge. Some 100 years ago, German physiologists were recommending approximately 145 g of protein per day for adults. This attitude was still prevalent during the First World War. Some observers have suggested that this contributed significantly to Germany losing that war, as the Germans persisted with traditional farming with large herds of sheep and cattle to supply this protein. The Allied blockade severely restricted food imports and there was insufficient dietary energy available from this traditional German farming. It is estimated that crops (e.g. cereals) grown on the same land would have produced at least six times the amount of human dietary energy. Fortunately, the British rapidly increased cereal production with the advent of the Second World War. By then it had also been realised that protein requirements for maintaining health were less than had previously been thought. The League of Nations Technical Committee on Nutrition (1936) recommended a daily intake of 1 g per 1 kg of body weight for adults (most adults will be within the range 50–90 kg). In 1973, a joint committee formed by the Food and Agriculture Organisation of the United Nations and the World Health Organisation suggested an 'average safe level of protein intake' for adults of 0.55 g per kg of body weight per day. Coincidental with this level of recommendation has arisen the possibility that a disease of malnutrition like **kwashiorkor**, which was thought to be primarily due to a lack of protein, may be due to a lack of food in general and not specifically to a lack of protein.

To maintain health at this level of protein intake it is wise to consume a variety of food and not rely significantly on one staple. This is the value of the many hundreds of crops which are used as food for man. Any deficiencies of nutrients in one food are usually compensated by higher levels in other foods. So William Cowper's thought that 'Variety's the very spice of life, that gives it all its flavour' still has relevance even after some 200 years.

One group which is grown worldwide and provides both variety and protein is the legumes.

Legumes

As with grasses previously, some confusion is sometimes apparent with the word legume(s).

A **legume** is a type of fruit, sometimes also called a **pod**, examples of which are the familiar pea and bean pods. It is formed from a single carpel, has seeds in a single row, and splits along both sutures to release the seeds.

This type of fruit is borne by a large group of agriculturally important plants, and so these plants are known collectively as legumes. This group of plants is second only in worldwide agricultural importance to the Gramineae. (The term legumes does not have the same taxonomic accuracy as Gramineae and reference works should be consulted for further information.)

Legumes are valuable for several reasons.

(a) Their seeds contain more protein than cereal grains. They frequently contain at least 20%, and usually more, protein (e.g. broad bean about 30%; soya bean about 35%).

Pulses (the edible seeds of legumes) are produced for human and livestock consumption. Whilst we are familiar with peas (*Pisum* spp.) beans (*Phaseolus* and *Vicia* spp.) and ground nuts/peanuts (*Arachis hypogaea*), there are many others, e.g. soy(a) bean (*Glycine max*); mung beans and cow peas (*Vigna* spp.); chick peas (*Cicer* spp.); lentils (*Lens esculenta*); pigeon peas (*Cajanus* spp. – human food in the tropics).

(b) The seeds contain a greater proportion of lysine than the other cereals (see Table 3.1).

They are therefore valuable as a complementary food to the cereals which are deficient in this amino acid. Therefore people existing on a staple cereal diet would be well advised to include pulses in their diet. They are usually easier and cheaper to produce than attempts to obtain meat or other animal protein. However, these pulses must be properly prepared as some contain toxic chemicals which only disappear after prolonged cooking, e.g. Lima beans; Locust beans.

(c) The agriculturally important legumes form **root nodules**. These are the site of a symbiotic relationship with bacteria (e.g. *Rhizobium* spp.) which allows the utilisation of **atmospheric nitrogen**. Not only is this obviously a major source of nitrogen for the plants' protein formation, but the residual effect is an increase in the nitrogen content of the soil, which benefits subsequent crops and thus enhances total yield from the soil.

(d) Some legumes are useful forage crops for livestock. They may be grown as pure stands, or possibly mixed with grasses. In the UK the main forage legumes are clovers (*Trifolium* spp.). They provide higher protein in the leaves compared with grasses, and the root nodules will provide residual nitrogen. This will even encourage grass growth if a mixed grass/legume sward is grown. A grass/clover mixture is the traditional 'seeds' of the Norfolk four-course rotation (roots – barley – seeds – wheat) (p. 28). There is a disadvantage in growing legumes which is related to their growing habit (refer back also to Gramineae, Chapter 1). Their growing points (meristems) are easily removed during grazing, and are particularly prone to damage during over-grazing. They are therefore easily killed, and, in a mixed sward, less able to compete with the grasses. If undergrazed, however, they may crowd out the grasses –

this even happens on badly managed household lawns! So precise management is called for if legumes are to prove advantageous for grazing.

Other legumes are also used for forage. Whereas the clovers seem to thrive more readily in the British climate, other legumes are favoured in drier areas. In general the legumes will put down a deeper rooting system than the typical fibrous root system of grasses which is often quite shallow. This deeper system allows the legumes to survive in the drier areas. For example, large areas of lucerne/alfalfa (*Medicago sativa*) are grown in the USA, Canada and Europe. Lupins, trefoils, vetches, sainfoin, etc. are used in limited amounts throughout the world.

(e) Some legume seeds are rich in oil, e.g. soya, and these oils can be extracted, refined and used as cooking oils, and in the manufacture of margarine, lubricants and soap.

Other important crops

There are areas of the world where plants producing underground storage organs are staples, e.g. **sweet potato** (*Ipomoea batatas*), various **yams** (*Dioscores* spp.), **cassava/manioc** (*Manihot esculenta*).

These are valuable mainly for the carbohydrate they contain but they are low in protein and alone do not provide sufficient protein for human requirements. They have the advantage that, if necessary, they can be left in the soil and lifted when required. Whilst the ravages of fire and rebellion may damage the visible shoots, the underground edible parts usually remain undamaged. Cereals would be completely destroyed by such actions. In some areas attempts to introduce what are thought to be potentially higher yielding crops, like the cereals rice or maize, have failed because these cereals need to be harvested at one time and then stored in specially constructed buildings. It is not always possible for countries to provide these harvesting or storage facilities.

The **potato** (*Solanum tuberosum*) is an interesting example of a crop which was a staple for many people in recent history, but which is now much less important. The reason for this change of emphasis is probably because the potato is basically a temperate area crop, and, in the main, poverty is nowadays more widespread in tropical than temperate areas. As stated earlier, it is poor populations that rely on a single or a few staples for their diet. However, in temperate areas the potato still supplies significant amounts of energy to people, and could be regarded as the world's leading vegetable crop, only being outranked by the three major cereals.

There are many other crops throughout the world which are not important for the carbohydrate or the protein that they contain, but more for their taste and minor nutrients (vitamins and minerals), e.g. fruits, salads, spices and beverages.

It must also be remembered that plants can provide useful **fibres**, e.g. cotton, which is the fibrous protuberances from the cells of the epidermis of the seeds of *Gossypium* spp. (the cotton seeds are pressed to release their oil, and the remaining squashed seeds form a valuable stockfeed); flax, the fibrous stem cells from *Linum usitatissimum*.

3 Farming systems

3.1 Introduction

Whilst it is not easy to define the word system clearly, it is even less easy to define farming system. In the context of this book, it means the organisations of various agricultural organisms and procedures to form a unified approach which produces something more than the separate parts could ever achieve alone. There are, however, a wide variety of farming systems and their different elements render classification difficult. For example, the traditional farming of the UK during the 18th and 19th centuries was based upon the **Norfolk four-course rotation**. It is rarely referred to as a farming system, yet it is one of the best examples. In it there was an integration of land, labour and a variety of livestock and crops into a very effective system. At the other extreme, reference is often made to **intensive systems**. In these only one type of crop or animal is husbanded, with only a little variation in the husbandry technique, and so there is very little integration possible. So the term farming system is a very loose defintion. Nevertheless, it is possible to trace an evolutionary path through these systems to see the development that has taken place. Evolutionary is perhaps not the most appropriate word, for some communities today still use systems which others abandoned some centuries ago and in some areas there is no, or very little, variation possible in the agriculture practised.

It was mentioned in the preface that thousands of years ago man lived as a hunter-gatherer. In this life-style, no attempt was made to influence the way in which crops or animals grew. Edible parts of plants were collected from where they grew naturally and wild animals were hunted and killed. Whilst people existed in small communities and had available large areas of land from which to hunt and gather, and assuming that a variety of plants and animals were consumed, it was nutritionally satisfactory.

It is important to realise that the social life of such communities also had an important bearing on the food collection system. Even though there was cooperation, all the people (or the members of a family group) had to be capable of performing all the tasks that life demanded – housebuilding, clothes making, defence, food preparation etc. There was no, or very little, **division of labour**, for this could only be successful in a settled community.

As the division of labour appeared and increased within a community, there began a situation which meant that someone was performing a task for the benefit of all the others. This was a full-time task and would not allow the individual sufficient time to collect or produce his own food. Even in this country, this situation was not completed until quite recently. It is important that we realise that others perform vital tasks for us, which we may never do, e.g. slaughter animals, build houses, weave cloth, extract metal and make a knife, etc.

Increasing numbers of **specialists**, coupled with the pressures of an increased population upon a finite area of land, meant that farming systems evolved which produced sufficient for the growers as well as an excess for sale (or barter) to feed the specialists. This represented a change of outlook, from farming for **subsistence** just to keep yourself and relatives alive, to farming for **profit**, producing excess for sale. This continues today. For example, in Britain less than 5% of the population produce about 50% of the food for over 50 million people.

3.2 Settled agriculture

Settled agriculture began to appear some 8000 to 10 000 years ago: maybe in one area but probably in several areas. The first problem with any form of settled agriculture was to eliminate the indigenous vegetation of the area and usually the trees were the main problem. A technique was used which is often referred to as **slash and burn**. The vegetation is cut down (slash) and then fired (burn). This is the easiest way of removing cumbersome trees (remember that this is before the production by domestication of draught animals capable of pulling fallen trees about, and that many tribes still practising this type of agriculture do not have draught animals in any great numbers, if at all).

The nutrients left behind from the burned vegetation represent valuable residues for subsequent plant growth. This is particularly true in the tropics where continuously warm temperatures and abundant moisture mean that there is little organic matter in the soil, for it is rapidly decomposed under these conditions.

The **fertility** of such soils is rapidly reduced after a few crops, and other areas have to be similarly cut and fired. Whilst some tribes must undoubtedly have left an area of land never to cultivate it again, others remained and cultivated several areas within easy reach of their settlements. A few years without cropping would allow fertility to return, yet the trees growing on the land would not be mature, and so recultivation would be easier than the initial slash and burn.

It is speculated that this type of slash and burn must have happened at least once on all the land currently under settled agriculture. This indicates one of the constant problems of agriculture. Agriculture can be regarded as an **artificial ecosystem**, constantly trying to return to the natural ecosystem and the natural climax vegetation. Whilst slash and burn occurs rarely in the UK today, it does happen, often under much protest. It should be remembered that many areas of this country were wooded until only a century or two ago when wars fought with wooden ships consumed vast areas of woodland, which often then went to cultivation rather than being returned to woodland.

A description by Roy A. Rappaport (1971) of the Tsembaga in New Guinea suggests a type of tropical agriculture involving slash and burn which must be very similar to the first tropical settled agriculture.

Whilst it is often speculated that man saved the seeds from plants and then sowed them at a later date to begin this type of agriculture, it must be realised that many **tropical crops** reproduce **asexually.** Yams, cassava, sweet potatoes and bananas all reproduce asexually and cultivation of this type of plant is comparatively easy. Often several different kinds of these crops are grown together on the same piece of land, all intermingled. This is common in the tropics on small farms and is sometimes referred to as **mixed cropping**. It should not be confused with **mixed farming** which is common in many countries and means having both crops and livestock on the same farm, though not usually on the same piece of land.

It is interesting that many crops of **temperate agriculture** are propagated **sexually**. This is probably because seeds with a low moisture content survive better in harsh winters than vegetative parts with a much higher moisture content.

This type of settled agriculture was initially only concerned with crop husbandry. Only when regular areas of grazing were established was it possible to exploit animals. There are still areas where animals follow the rains and, therefore, new vegetation.

Various agricultural systems evolved in temperate areas from this initial settled agriculture. Much tropical agriculture remained (and still remains) at the same stage until colonial influences within the last two hundred years.

3.3 Crop rotations

By Roman/Saxon times, the agriculture around settlements had evolved into a more organised system – the simplest **rotations** had developed. Initially about half the land would either be sown to a winter cereal (in autumn) or a spring-sown cereal. The other half would be left in fallow, being cultivated at regular intervals to destroy weeds. Eventually a gradual change to one third winter cereal, one third spring cereal and one third fallow occurred. There was little apparent loss in fertility from having two succeeding crops and one fallow, yet it allowed an increase in cereal production. So the **three-field** rotation had evolved.

Figure 3.1

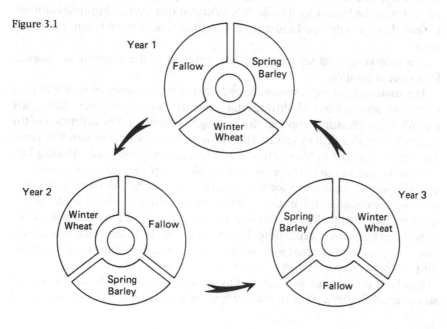

Notice that the winter cereal would be sown in the land that had previously been fallow. The winter cereal would be harvested in the following late summer and the land ploughed during autumn and winter ready for the spring sowing of the spring cereal. There would be insufficient time to plough the land and cultivate it after harvest to allow an autumn sowing of a winter cereal.

Remember also that the land was in strips and that cooperation was important for the cultivation of these strips, for individuals did not all own the required oxen and cultivation implements – it had to be a cooperative effort.

At this time cereals were the only major crops. Other minor crops would be grown in much smaller quantities on land adjacent to the households. Sheep and cattle would roam the common land and other uncultivated areas. Pigs were still the hunted wild boar.

Such situations continued until the 18th century when other crops became agriculturally important, particularly **root crops**, like turnips, and improved **grasses** and **clovers** for better grassland. Under the impetus of this development and as a consequence of **enclosures**, which (whatever their social implications) allowed fences/hedges/walls to be erected to protect growing crops from animals, there developed a **four-course rotation** from the previous three-field rotation. In this rotation, cereals alternated with roots or temporary grassland. This rotation became known as the **Norfolk four-course rotation** because of the influence of that county and some of its farmers upon this agricultural development. It is often summarised as 'roots – barley – seeds – wheat'. Whilst the system was **labour-intensive** (which was one reason for its disappearance) it was probably the most successful rotation and agricultural system ever devised. It was skilfully integrated so that different activities did not clash, but followed each other consecutively through the year; it maintained (and even raised) the fertility of the soil; it controlled weeds successfully and provided sufficient food to feed the increasing number of people in towns who were the mainstay of the Industrial Revolution. Without this 'Agricultural Revolution' to feed these people, the Industrial Revolution would have taken a different path.

The following diagram represents the rotation of the crops in the Norfolk four-course rotation.

The **roots** (e.g. turnips, swedes, mangolds) would be sown in spring in land that had been ploughed during the previous autumn/winter. Roots are regarded as a 'cleaning crop' but this is a slight misnomer. The invention of the horse-drawn seed drill by Jethro Tull allowed these roots to be sown in rows. They were thus the only crop that could be hand-weeded easily. Hoeing kills the weeds and so really roots are a 'cleanable crop'. By autumn these roots would be grown sufficiently for feeding to livestock during the winter. Turnips and swedes could be left in the fields. They are fairly hardy to frost, swedes more so than turnips. They could be carted to the buildings to feed cattle, or left to be eaten *in situ* by sheep. Mangolds are much less resistant to frost, and were stored in long piles covered with straw and earth called **clamps** – the same as for potatoes. They would then be used for late winter feed.

The land would then be ploughed as soon as possible ready for the **spring-sown cereal** (barley). After the barley had grown to about 10 to 12 cm, 'seeds'

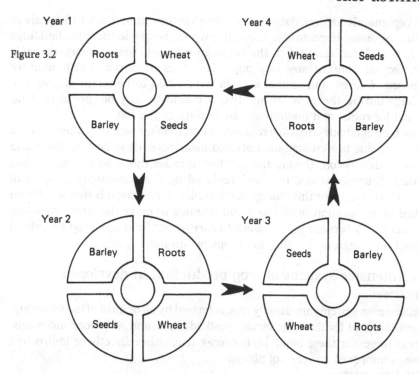

Figure 3.2

Year 1: Roots, Wheat, Barley, Seeds

Year 2: Barley, Roots, Seeds, Wheat

Year 3: Seeds, Barley, Wheat, Roots

Year 4: Wheat, Seeds, Roots, Barley

would be sown by hand (broadcast) into the crop. The 'seeds' is a term representing a mixture of grass and clover seeds. If sown at the same time as the barley, they would compete with the barley and reduce its yield. Providing the timing of the sowing was correct, by the time the barley was harvested in the late summer, the grass/clover would be a few centimetres high. This would then give a feed for livestock in autumn and would then be used as grazing or for haymaking in the next spring and summer. In late summer it would be ploughed in. This is an example of a grassland lasting for a short time, in this case just a year. These temporary grasslands are often called **leys**, and are still very important in modern agriculture. The clovers in the ley would have added to the nitrogen content and hence the fertility of the soil because of the root nodules present. After cultivation the winter wheat would be sown, harvested the next year, and the land prepared for the root crop sowing the next spring. So the rotation would have gone full cycle.

Modifications to this rotation included the growing of potatoes, and later sugar beet and kale at the 'roots' stage of the rotation. Extensions of the cereals by a year or two also occurred and foreshadowed the modern continuous cereal growing. Some areas never really adopted this rotation. The area around the Wash had very fertile soil (p. 69 *et seq.*), and usually crops were grown and very few animals kept. On these farms there would be two or three years of cereals, followed by root crops, possibly for sale, e.g. potatoes.

The animals kept under this system were also important. Sheep represented an important income from both mutton/lamb and wool. The winter feeding allowed the development of larger, faster growing breeds (see p. 62). Cattle

29

were kept mainly for beef, dairying only developing within the last 150 years to its current major importance. The cattle would be inside the farm buildings during the winter being fed on the home-grown root crops and hay, and also being fed on concentrates (see page 51). This produced a rich manure composed of straw and animal excreta which had to be removed from the buildings during the summer months. It would usually be spread on land destined for roots, and maintained the fertility of the soil.

This type of labour-intensive rotation began to falter seriously after the First World War due to a combination of circumstances such as cost of labour and economic depression. During the last few decades this type of rotation has virtually disappeared and has been replaced by the continuous growing of crops. The reasons for this change are complex. One reason is that a rotation implied an integration of all crops and animals to make the farm profitable, whereas it has become much easier to earn profit from one crop or animal without an integration with other crops or animals.

3.4 Intensive systems of crop production in developed countries

Intensive crop systems are usually characterised by all or most of the following:
– high levels of fertilizer, herbicide, pesticide use, and advanced mechanisation (together these imply high energy use, either directly or indirectly);
– use of improved varieties of plants;
– low labour input;
– maximum yield from minimum land area;
– attempts to maximise profits.

Crops are usually grown as a **monoculture** in developed countries. This means that only one crop is grown and husbanded on one piece of land. This has made it much easier to intensify crop production, and incorporate the characteristics mentioned above. Some of the factors which have allowed intensive crop systems to develop have also encouraged **continuous cropping**. Continuous cropping is the growing of the same crop year after year on the same piece of land as opposed to crop rotation. This continuous cropping has arisen as it is a profitable method because of the employment of expensive machinery over large areas and because advances in the use of herbicides, fungicides and insecticides have made rotation less necessary. Many of these factors are also related to **increased yield**, and so, in several instances, continuous cropping and increased production have occurred together.

Several crops are now grown worldwide in an intensive manner. The continuous growing of maize and wheat in northern America was the first example of this. Cereal growing in the UK during the last three or four decades also provides a good illustration. It has been mentioned earlier that UK government policies encouraged cereal production after the Second World War. A system of guaranteed prices was announced in advance, which allowed farmers to plan ahead with a much greater measure of success and security than the previous 'unknown market forces' price policy.

It is an incontrovertible fact that yields of cereals have risen dramatically in the last few decades. This is shown by Figure 3.3.

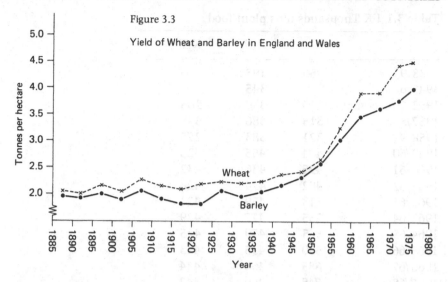

Figure 3.3

Yield of Wheat and Barley in England and Wales

Several interesting points arise from these figures. Plotting both wheat and barley on the same graph shows how, in general, the figures move together. If one surmises that similar management techniques are applied to each crop, then this collateral movement of yield is an expression of variations over which the farmer had no (or very little) control, such as climate, genotype of the crop and variations in pests and disease. Whilst variations in the climate would have occurred during the period shown on the graph, these variations are small compared with the increase in yield that occurred from the 1940s onward. Therefore other factors (not climatic) must have contributed to this increase in yield.

In the 1950s **new varieties** of cereals appeared which had inherently better yield potential than previous varieties. Probably two of the most important at that time were Cappelle Desprez (wheat) and Proctor (barley), from France and the Plant Breeding Institute, Cambridge, respectively. More recent, superior varieties have now replaced these.

Also, importantly, **herbicides** were becoming commercially available. Herbicides developed as the result of research work on naturally-occurring auxins and apical dominance. The first commercially available synthetic auxin was the herbicide 2.4-D. This was widely used for the elimination of broad-leaved weeds in cereal (and grass) fields.

Also at this time, **inorganic fertilizers** were more readily available. Table 3.1 shows that whilst phosphorus and potassium have reached a level at which they have remained for years, nitrogen use is still increasing. For further information on fertilizers see Finch (1977).

The use of these new varieties, herbicides, pesticides, and increasing amounts of fertilizer all contributed to this rise in yield. The yield figures are an arithmetical mean derived by dividing total yield by total area sown. They do not show the maximum or minimum achieved by individual farmers. This variation is important because another aspect emerges from consideration of the area of cereals grown.

31

Table 3.1 UK Thousands tons plant food

	N	P_2O_5	K_2O
1938/9	50	195	40
1945/6		345	
1952/5 (average)	245	350	265
1957/8	315	386	348
1958/9	321	383	375
1959/60	404	455	426
1960/61	425	436	432
1961/62	482	474	426
1962/63	513	459	421
1963/64	555	477	429
1964/65	565	479	425
1965/66	590	435	413
1966/67	685	460	434
1967/68	748	469	442
1968/69	769	460	440
1969/70	795	470	419
1970/71	894	504	441
1971/72	919	473	416
1972/73	947	482	417
1973/74	980	449	421
1974/75	980	393	377
1975/76	1059	404	398
1976/77	1093	406	409
1977/78	1155	410	412
1978/79	1186	416	416
1979/80	1286	440	444

Source: The Fertilizer Manufacturers Association Ltd.
(From 1969/70 figures are in metric tonnes (1000 kg) which introduces an error of approx. $1\frac{1}{2}\%$.)

Figure 3.4 shows that since the 1930s the area of cereals grown has increased. Whilst wheat has shown some increase, the area of barley grown has approximately doubled.

Some of this is undoubtedly due to barley being grown where oats were grown previously (see declining figures of oats). However, this would account for only about one half of the increased barley area. This means that as the barley area increased, it was increasingly being grown on land which was previously considered marginal for barley. **Marginality** is a concept of much discussion, but, roughly, it is the close relationship between output (yield) and input (fertilizers, machinery, labour, etc.) on land of less than average fertility. This implies that the barley produced would only just pay for these inputs (on better, more fertile land, much greater profits would be expected). The new varieties of barley were capable of giving a better yield even on this marginal land, assuming the same level of inputs. With the price incentive previously

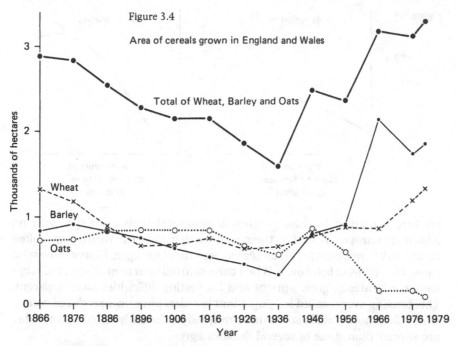

Figure 3.4

Area of cereals grown in England and Wales

Total of Wheat, Barley and Oats

Wheat

Barley

Oats

Thousands of hectares

Year

mentioned, it was now possible to obtain a reasonable profit on this marginal land. Thus more and more land was sown to barley (see increasing area figures). However, much of this was even more marginal. Whilst yield was adequate, it inevitably depressed to some extent the average British yield. This is one reason for the diminution of yield increase in recent years.

The continuous cropping that is practised means that pests, weeds and fungal diseases are likely to increase because they more readily find suitable conditions. Also, an increased area of crop grown means that, on average, fields of the same crop are nearer to each other than when a lesser area is grown. This proximity and prevalence means that the growth and the spread, particularly of fungal diseases and insect pests, is much greater. These conditions allow build-up and much larger populations per season than were previously apparent.

Measures taken to prevent these organisms, e.g. fungicides and insecticides, are never completely successful, for it is likely that some organisms will be **resistant** to these measures and survive. These survivors will probably be genetically different from the others that were killed. With little or no competition for food, these new races will soon multiply and become a major part of the population of that organism. During the last two decades, several pests and diseases have become very prevalent (e.g. smut and rust fungal diseases, and aphids) on cereals. These were rarities before the 1950s.

Plant breeders have produced new varieties with resistance to some of these fungal diseases. Unfortunately their efforts are often short-lived due to the evolution of new races of fungi.

Another phenomenon which has had some effect on yield is **lodging**. When increased amounts of nitrogenous fertilizers are applied to cereal crops, they

Figure 3.5

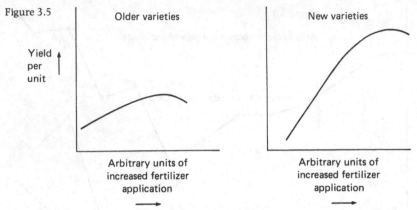

produce a marked increase in growth compared with a control with no additional nitrogen. However, the stem growth in the early varieties was often quite weak in response to this increased amount of nitrogen. Heavy rains soon caused the crops to bend over. When this occurred near to maturity, the crops were rarely able to grow upright and harvesting difficulties were apparent. This bending over is called lodging. Plant breeders have now produced cereals with much shorter, stiffer stems that rarely lodge. Certainly the cereals of today are shorter than those of several decades ago.

Another comparison with previous varieties reveals a difference in uniformity. To obtain **consistency of yield**, plant breeders have produced cereal varieties with virtually identical genetic make-up within each variety. This means also that, across a field of one variety, all the flowering heads of the plants will be at the same height due to this genetical uniformity. This was never apparent in the older varieties. They were less genetically uniform (some were varieties selected and kept by the farmer). This shows in photographs of corn fields from the past for the heads of the cereals are at a variety of heights. This genetic variability had advantages. The different plants within a variety were probably differentially susceptible to fungal diseases and so not all the crop would succumb to a disease, and disease would spread less rapidly and rarely be the problem that it is with continuous cropping and genetic uniformity.

The response to nitrogenous fertilizers that was mentioned is another reason why the new cereal varieties were readily accepted. The varieties of the 1950s had a good response to fertilizer. Some of the older varieties, and also oats, were much less responsive to nitrogenous fertilizers. This difference is indicated in Figure 3.5.

With both of the curves there is a decreasing response at the higher levels of fertilizer input, and in these circumstances the cost of extra input is not met by the small extra yield obtained. The response often works in the opposite direction. If very little or no fertilizer is added, the yield may be poor (in some cases even poorer than the old varieties that were replaced). Whilst this does not usually happen in the UK, in developing countries farmers can rarely afford much fertilizer. If plant breeders supply new varieties of cereal with a high response to fertilizer to areas where little fertilizer can be afforded, then there is the distinct possibility that these new varieties, given no fertilizer, may not yield as well as the indigenous varieties, even though their potential is much greater. It is unfortunate that the literature suggests that this may have happened.

One of the most recent developments in continuous cereal growing is **direct drilling**.

After harvest, the stubble (remains of the cereal) is sprayed with a herbicide such as Paraquat. This kills all foliage with which it comes into contact and so kills all the weeds that are growing. Paraquat is not a persistent herbicide, being rendered inert in the soil.

After the killing of the weeds and cereal crop remains, the next cereal is sown using a seed drill which is capable of breaking through the undisturbed soil and planting the seeds. These cereal seeds then germinate and emerge with little or no weed competition.

This direct drilling system represents a saving in energy. The fossil fuel used to produce the herbicide is less than would be required in the traditional method of ploughing to invert the soil to kill the weeds and then cultivating to prepare a seed bed. In general, the manufacture of herbicides uses much less fossil fuel energy than mechanical weeding operations (Green, 1979).

4 Conservation of grass

4.1 Introduction

There are many areas of the world where the growth of grass is seasonal. In temperate areas this **seasonality** is mainly due to variations in **temperature**, and ample growth only occurs in the warmer months. In more equatorial areas, temperatures are higher and seasonality of the **rains** is the main factor controlling grass growth; the grass grows during the rainy season and virtually ceases growing during the dry season. Farmers all over the world keep grazing animals throughout the year and therefore food is needed for these animals during the non-productive months. Whilst some nomadic peoples may move their animals to follow the rains and thus find growing vegetation, this is not feasible under settled agriculture. So grass is conserved (cut and preserved) from the productive season to supply the bulk of this required food.

Grass, like all leafy crops, will deteriorate rapidly when cut. This is due to both internal and external factors. Internally, the dying cells autolyse, and respiration continues until the moisture is removed. Externally, micro-organisms act upon the cells.

For the crop to be conserved, this rapid deterioration must be halted. The methods theoretically available are the same as those used to preserve similar types of foods for human consumption. However, whilst deep freezing is widely used to preserve human food, there are currently only two methods practicable on the farm: **dehydration** and **an acidic environment**.

The product of dehydration is called **hay** (do not confuse with straw, the flowering stems from cereals). The process takes place in the field and utilises direct solar energy to remove the moisture. The completion of the process is sometimes assisted, when the part-made hay has been collected into buildings, by electric fans, and possibly some artificial heat (**barn drying**). A very small amount of hay is conserved using fossil fuels for the whole process (**green crop drying**).

An acidic environment produces **silage**. The cut crop is transported to a **silo** close to the farm buildings and micro-organisms produce **lactic acid** to conserve the crop. This is the equivalent of pickling.

Whilst grass is the crop mainly conserved, it is possible to use other crops. Many farmers grow legumes, e.g. red/white clovers, mixed with the grasses. The legumes have a higher protein content than the grasses. However, their leaves tend to shatter and be lost during haymaking, and they are often more difficult to **ensile** (turn into silage). Sometimes pure stands of legumes are grown. In the UK, maize has recently gained some popularity, particularly in southern England. In the UK maize only achieves maximum vegetative yield in late summer/early autumn. This can be quite considerable, and the maize is then ensiled. It is not made into hay as it has wide stems, which make water extraction difficult, particularly in autumn.

4.2 Haymaking

Grass, at around 70–80% moisture content, has to be dried to approximately 15% for successful storage as hay. The natural drying processes of the sun and wind will remove most of this moisture, the remainder being lost during the initial storage period. In the UK most hay will be made during June/early July. At this time the grass is less nutritious than at earlier periods (see later). However, the yield is high, and haymaking operations are expensive and have to be performed whether the crop is 'light' or 'heavy'. Also, most people consider that the weather is better at this time, rather than earlier.

The standing crop is cut with a mower which has a reciprocating knife on a cutter-bar (see Figure 4.1).

This leaves the fallen crop in rows called **swath(e)s**. On fine days the grass will soon lose moisture and wilt. Machines which lift, turn and fluff-up these swathes are now used to obtain maximum exposure to sun and wind. Whilst there has to be a compromise between the use of expensive machinery and fuel, and low-cost hay, the best quality hay is usually that which is made rapidly because the intra-cellular respiration is reduced rapidly, leaving maximum cell contents for food value. Also, there is less likelihood of loss of soluble nutrients by leaching due to rain. Two/three day periods of sunshine are far more prevalent than four/five day periods of sunshine.

Figure 4.1

When at a moisture content of very approximately 30%, most hay in the UK will be **baled**. An estimate of the optimum time to bale will be judged by the farmer by smell, feel and experience – not by the use of moisture meters. Machines called balers pick up the hay from the field, compress and tie the hay into bales. These will either be cuboid structures (about $1\,m \times \frac{1}{2}\,m \times \frac{1}{2}\,m$) or, a more recent development, large cylinders weighing about $\frac{1}{2}-\frac{3}{4}$ tonne (about 10–15 cwt), usually moved by tractor.

In store, further moisture will be lost down to about 15% moisture content. This occurs because the very slight moisture still present will allow some respiration to continue. This causes a little heat to be produced which helps remove moisture, probably by slight air movement due to convection currents.

For the farmer, there is often the compromise between storing at too high a moisture content, or leaving in the field and risking rain spoilage. Storing at too a high moisture content will allow more respiration. Overheating will then occur which lowers the food value (denatures proteins; reduces palatability) and may even result in so high a temperature as to cause spontaneous combustion. If the weather seems likely to be fine, then there is the temptation for the farmer to leave the crop until it is much drier. However, a crop that is too dry can present problems. When a very dry crop is moved there is the likely possibility that crop losses will occur, for leafy parts are brittle and will shatter into small pieces and thus be lost. This is a particular problem for legumes.

In the wetter parts of the country, e.g. the Lake District and parts of Scotland and Wales, it is often impossible to make hay in one single dry period. Partially made hay is then sometimes made into small **haycocks** in the field. These afford some protection against the rain, and in sunny periods they will continue to dry out, especially if made over tripod or quadpod structures which allow air to circulate inside as well.

Whilst haymaking seems to be a less critical process than silage-making, and may produce a highly aromatic and 'attractive' product, its nutritive value is often over-estimated, and increased labour, machinery and fuel costs, coupled with 'invisible' losses (leaf shatter, final product of lower nutritive value than silage) may make its food content more expensive than has been previously realised. So nowadays, for a more nutritious product, many farmers, particularly those with high yielding dairy cows, make silage.

4.3 Silage

The process is sometimes called ensiling, the product silage and the site a silo. Many silos are simple constructions (see Figure 4.2).

Silos have often been made of second-hand timber, e.g. railway sleepers, lined with impervious sheeting. Some farmers use tower silos. These are usually linked to adjacent feeding areas, and the whole system is often highly mechanised – the touch of a button feeds the animals. The product is often more easily made because tower silos can often be hermetically sealed. The drawback is the usual financial one.

The grass has to be cut into short pieces to allow for easier compaction

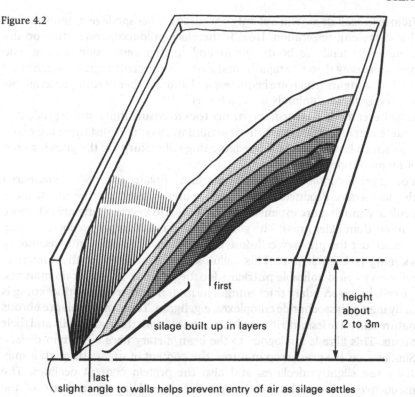

Figure 4.2

first

silage built up in layers

height
about
2 to 3m

last

slight angle to walls helps prevent entry of air as silage settles

within the silo. This is necessary for successful preservation (see below). To achieve this short grass, some farmers use a **forage harvester** which directly cuts and chops the crop. This is then blown into trailers pulled behind the machine and subsequently transported to the silo. This can often be satisfactory, but may present difficulties when the crop has a high moisture content. A compromise is to cut the grass into swathes with the ordinary reciprocating knife mower, allow the grass to wilt for several hours and then pick up and chop with the forage harvester.

The cut crop is placed in the silo in layers using a tractor with a large buckrake (see Figure 4.3). The weight of the tractor is used to compact the grass, and finally plastic sheet to seal the silo.

The preservation of the grass is brought about by the production of an acidic environment. Under **anaerobic conditions**, various bacteria will ferment sugars to **lactic acid**. This lactic acid lowers the pH (down to about pH 4–4.5) which prevents any further microbial despoilation and so effectively **pickles** the grass. The chopping of the crop ensures a ready availability of cell contents, and hence sugars, all over the surface of the crop for the bacteria to metabolise into lactic acid, as well as allowing ready compaction to exclude air (anaerobic conditions). If the moisture content of the crop is high, then these sugars are less concentrated. Similarly the lactic acid produced will be less concentrated, and there may be insufficient to lower the pH enough to prevent undesirable organisms from spoiling the grass. Spoilage produces unpleasantly odoured compounds including butyric acid, and proteins are degraded and may putrefy. Ensiling is often encouraged by acidic additives.

Failure to exclude air will not only encourage other spoilage organisms, but will also prolong respiration. Heat is therefore produced, convection occurs and more air tends to be drawn in and further respiration occurs. The temperature may then rise rapidly, and also cause protein degradation, as well as the loss of sugars which are being respired and will therefore not be available in the silage for the animals to which it is fed.

Whichever method is used, the farmer tries to ensure that a stable product of adequate nutritive value is produced at minimum cost. Whilst there are many factors which have to be considered, the stage of maturity of the growing crop is of major importance.

Young grass contains a higher proportion of water (over 80%) compared with older grass. Products with a high water content can be much more difficult and much more expensive to conserve. However, young grass is more nutritious than older grass. The young grass has a high proportion of young cells with just the **primary cellulose cell wall**, and few cells with **secondary thickening** (sclerenchyma). This cellulose is metabolised in the rumen of ruminants to yield valuable nutrients for the animals. As the grass matures, more cells have secondary thickening. The material of secondary thickening is usually inert polysaccharide complexes, e.g. **lignin**. These materials are fibrous in nature and have less nutritive value compared with the young cells and their contents. This fibre is analogous to the bran/dietary fibre of human diets.

Similarly, as the grass crop matures, the content of sugars (though it may initially rise slightly) declines, and also the protein content declines. The consequence of these changes is that the **digestibility** (a measure of the nutritive value of the food to the animal) decreases as the crop matures.

Figure 4.3

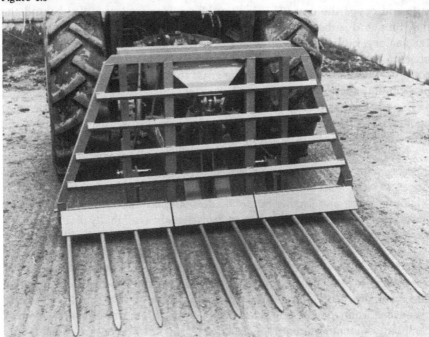

Running counter to this is the yield. The total yield (measured as dry matter) increases as the crop grows, and reaches a peak at or just after flowering. However, this is an increasing yield of progressively less and less digestible material. So a compromise has to be made between yield and digestibility. Farmers try to make silage at around 65% digestibility. This is a little before the maximum yield of digestible organic matter occurs. Experiments show that animals will eat more of the more digestible materials, and less of the less digestible materials. So the figure of 65% digestibility represents the approximate optimum between yield and acceptability. Whilst there is variation both within and between varieties and species, 65% digestibility occurs at about the time that the flowering head is about to emerge from the surrounding leaves and is visible without opening the leaves.

Figure 4.4 summarises these changes in the crop.

Figure 4.4

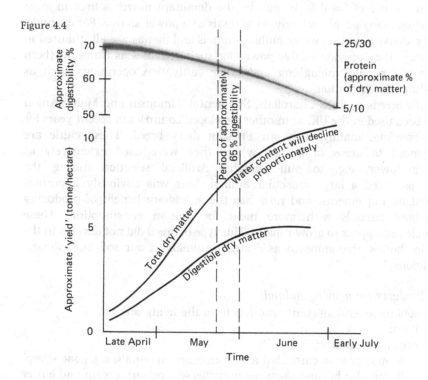

5 Livestock

5.1 Introduction
Animals are useful to man in several ways.

(a) Power
Whilst the use of fossil fuels may be the dominant power source in most countries, many people still rely on animals as a power source. For example, horses, donkeys, cattle, water buffalo, camels and llamas are all still used in this way. They provide motive power for such activities as transport (both goods and people), ploughing and other cultivation operations such as threshing and irrigation.

Cattle breeds such as Charollais, Simmental, Limousin and Maine-Anjou have been used in the UK, as in other developed countries in recent years for cross-breeding, mainly with our Friesian dairy breed. These cattle are indigenous to areas of Europe where they were used extensively as motive power, e.g. to pull ploughs. Artificial selection during the years produced a large muscular animal. This was obviously important for pulling implements, and now has the subsidiary benefit of producing cross-bred animals with more meat for human consumption. These animals also appear to grow rapidly. This type of breed did not develop in the UK, for horses predominated as draught animals in our soil and climatic conditions.

(b) Products from living animals
Man obtains several different products from the living animal.
 (i) Food:
      eggs from poultry;
      milk, mainly from cattle, but also from other mammals, e.g. goat, sheep. Milk can also be manufactured into cheese, yoghurt, cream and butter and these are products which have a much longer storage life than the untreated liquid milk.
 (ii) Other:
      wool, mainly from sheep, but also from goats, alpaca, rabbits, camel; dung, whilst this is useful as a fertilizer, its organic content means that for some people it is important as a fuel. It is also useful as a building material, either when dried into bricks, or mixed with soil and daubed on walls.

(c) Products from dead animals

Most animals are **edible**. Reports on the meat for sale in the markets in some countries indicate that our zoological gardens contain many potentially edible animals, e.g. rodents, antelopes, bats, anteaters, carnivores, primates, birds, reptiles and various invertebrates. Beef in Asian areas is often water buffalo, being apparently difficult to distinguish from 'cattle' beef.

The **skin** of the animal provides hides. Treatment, including removing the fibre, produces leather for goods and clothing.

Animal products which are not edible or wearable are also useful, after suitable treatment, as glues, fertilizers and other products.

(d) Other uses – for example, religious festivals

5.2 Animals for food

Despite this wide range of uses of animals to man, the vast majority of them are kept to provide food. It is in this context that there have been most advances in livestock keeping during the centuries, and it is in this context that the most familiar farm animals will be discussed.

The value of farm animals as food

The food produced by animals is of a **high quality**. Note was made in Chapter 1 p. 15 of the use of egg as a reference protein with which to compare all other proteins. **Milk** has long been recommended as a 'whole' food. All **meats** (i.e. all animal muscle tissues) contain in more than adequate quantities all the amino acids required by man.

There are large areas of the earth's surface where the cultivation of plants which could be eaten directly by man is either impossible or economically impracticable. This is caused by such factors as inaccessibility (e.g. slopes which cannot be ploughed); thin, poor soil; insufficiency of water, light and temperature. Under these circumstances the consumption of vegetation by animals represents the only viable form of agriculture, and adds to the total food available to man. Any discussion about animals eating food that could feed man just does not apply in these circumstances. It must also be remembered that these **livestock only** areas are not just in other countries. The tops of all the British hills and mountains will only support the indigenous grasses (apart from trees, which are the subject of a different debate). Under these circumstances, sheep, and possibly some cattle, represent the only possible agricultural enterprise, and their meat a valuable addition to our diet.

Man's need for food is continuous. In many areas, plants only provide food for part of the year. Whilst recent technological advances available in developed countries mean that many foods can now be stored for many months, this has not always been so, and is not a technique readily available in other countries.

Animals, therefore, act as a reservoir or intermediate store of nutrients, growing when food is plentiful, and being available as food for man in the leaner seasons.

Even this is a comparatively recent development in temperate areas, for, prior to the Agricultural Revolution of the 18th century, little food was available even to keep breeding and draught animals alive during the winter months. Reports suggest that some animals were so weak after a winter with little food that they had to be lifted from the buildings in spring into the fields to eat the newly-grown grass.

Developments of agricultural techniques in the 18th century made available winter feed for livestock. This was supplied particularly by turnips, a winter-hardy root crop, and by hay crops from the improved pastures. Man assumed responsibility for the winter survival of his stock, and selection produced larger and faster growing animals. The previous animals were somewhat like the present-day hill sheep breeds, which are best suited for winter survival: small, with a slow growth-rate and with large fat reserves put on during the summer to help survival during the winter.

Modern agriculture in developed countries has built upon these changes instigated some two centuries ago and now attempts to provide ideal food and environmental conditions for all its livestock so as to maximise food production.

The descriptions of livestock and, more particularly, the associated husbandry techniques, can only represent a general consensus view. It could be said, with little fear of contradiction, that there are as many ways of looking after farm animals as there are farmers looking after them. Each farm is unique. Someone, somewhere will be successfully farming in a way not described here.

It is probably no accident that the four important groups of agricultural animals (**sheep, cattle, pigs, poultry**) are **warm-blooded** (homoiothermic). This term implies that they maintain their body at a roughly constant temperature even though the environmental temperature may alter. They are therefore capable of surviving in a wide variety of environments. To maintain this constant internal temperature, they are capable of producing heat energy from within themselves by catabolism. They are therefore **endothermic**. Experiments with farm animals suggest that there is an inverse correlation between external temperature (up to about 30 °C) and food intake, for as the temperature rises, food intake decreases and vice versa. This suggests that the animals need to convert more of the food that they eat into heat energy to maintain their body temperature in a cold environment compared with a warmer environment. If the animal is already eating to capacity, less food will be available for production (meat, milk, eggs) and yield will drop or growth slow down. Thus for the economic production of some farm animals (most notably poultry) there is probably a balance point between the cost of providing a controlled environment and the cost of food for the animal. If this balance is not achieved then food costs or fuel costs (for the controlled environment) may be excessive. Unfortunately this is not always easy to determine as costs alter continuously.

In the previous sections on crops, the provision of adequate carbohydrate and protein for humans was considered. For the efficient production of most farm animals similar considerations can be used in formulating appropriate rations for them. It is possible to consult tables of nutrient contents and

Figure 5.1   Ruminant Digestive System

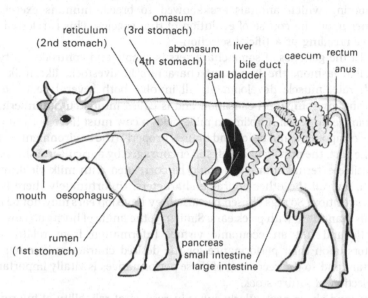

digestibility of foodstuffs for livestock, and advice is available from government bodies and feedstuffs manufacturers.

Sheep and cattle are **ruminants**, whilst pigs and poultry are **non-ruminants**. The plant material that the ruminants consume (mainly grass) consists of cells with a cell wall characteristically described as being made of cellulose. It does contain cellulose, but also other polysaccharides and associated complexes. The ruminant is able to eat large quantities of herbage relatively quickly, with little digestion, and store it in the rumen (see Figure 5.1). Later this food is regurgitated and then masticated (rumination or 'chewing the cud'). There is probably a considerable element of truth in the adage that cattle spend eight hours walking about eating, eight hours sitting down ruminating, and eight hours lying down asleep, per day. Rumination is important because it helps physically to break open the cells to release the cell contents for digestion. Note that rumination occurs in the mouth, and not in the rumen. The rumen is the site of fermentation, for apart from its storage function, the rumen is also important because of the microflora it contains. Whilst the cellulose complex is virtually indigestible in humans, the microflora (bacteria and protozoa) of the rumen are able to digest cellulose. The results of cellulose digestion by the microflora includes short chain fatty acids which are absorbed and metabolised, and gases such as carbon dioxide and methane (eructation has to occur regularly in ruminants or 'bloat' occurs).

The pig is a non-ruminant and has a digestive system that is very similar to humans. It is likewise an omnivore, though usually fed as a herbivore. Cellulose is digested by microflora in the small intestine and caecum.

Poultry, as birds, have a somewhat different digestive system which is discussed later.

45

5.3 Breeding of livestock

By choosing which animals are allowed to breed, humans exercise an influence upon the course of evolution of that species. This is referred to as **selective breeding** or **artificial selection.**

Whilst the genetics of some characters is simple and controlled in typical Mendelian fashion, the important characters in livestock, like milk yield, growth rate, muscle development, all involve both several genes *and* an important environmental causation and is called **multifactorial inheritance.** For example, to obtain maximum milk yield, a cow must have both the right genetic component, and be fed and housed properly (the environmental effect).

In the past, the prevalent belief, still encouraged by agricultural shows, was that visible external characters could be correlated with milk yield, muscle growth, and all the other desirable characters. Unfortunately there is little such correlation. So pigs chosen because they have lop ears may not be better animals than those with prick ears. Similarly, the angle of horns on cows is an aesthetic and not an economic virtue. Information from relatives and ancestors upon their performance for the desired characters is much more important and so the recording of these performances is vitally important for the selection of future stock.

Cattle and sheep particularly suffer in respect of reliability of information because of the relatively small numbers of offspring and the long generation and maturation time. However, artificial insemination has helped the situation in cattle, because sperm from bulls can be saved until worth is proven.

The most notable advances in the selection and evolution of breeds and new lines has taken place in poultry, for under modern techniques they have both the fastest generation time and the greatest number of offspring. This has allowed the relatively rapid selection of animals which are prolific, have good food conversion ratios, and also fast growth.

5.4 Cattle

In many tropical areas, cattle are often only useful as religious symbols, or as a power source. As such they are economic liabilities, consuming much food. In other countries they have two main purposes, to provide meat (beef) and milk. Artificial selection in Britain during the last two centuries has tended to emphasise these different purposes. There arose breeds which were primarily kept for **meat production**, like the Aberdeen Angus in the north of the country, and the Hereford in the south-west. These animals were traditionally reared under a system that is often referred to as a **single-suckler herd**. Each cow would produce a calf, usually in spring, which, with the initial impetus of the milk from the cow and the grass pastures of the summer months, eventually grew (in 2–3 years) into a large, stocky, well-fleshed animal. Figure 5.2 shows the typical cuboid shape of these meat animals. Some beef is still produced in this way.

Similarly, there arose breeds specifically for **milk production**, like the Ayrshire, the Guernsey and the Friesian. Cows were selected for their milk yield and through successive generations, roughly on the principle of

Figure 5.2   Hereford bull – a typical beef breed

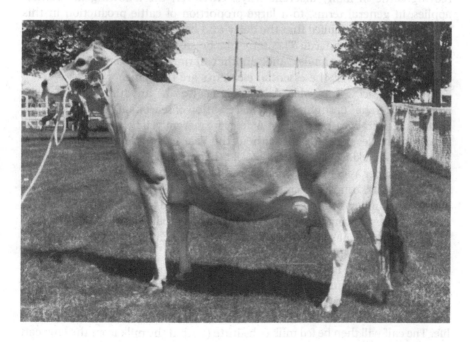

Figure 5.3   Jersey cow – a typical dairy breed

'bigger is better', the size of the udders of cows became larger. As little heed was paid to any beef potential, these animals had little extra flesh and were more triangular in shape (narrowing to the head), and often the pelvic bones were prominent as in Figure 5.3.

In some areas, the distinction between these two types was not clear cut and there arose breeds referred to as **dual-purpose**, e.g. Dairy Shorthorn, Welsh Black. These breeds were suited to areas where conditions were less favourable and they were able to give an adequate milk supply (for sale) as well as offspring that could be reared for beef. Their detractors argued that they fulfilled neither purpose with any measure of success, and certainly they have hardly featured in recent developments.

During the last few decades, coincident with the ready availability of quality of feed and with advances in knowledge, there has been a merger of the two functions so that most of the beef produced in this country arises as offspring of dairy cows, usually Friesians. This has arisen from an awareness that the offspring of the Friesian cow (both pure Friesian, and crossed to a beef breed, e.g. Hereford or Continentals) has potential for rapid growth as a beef animal. Previous ages never considered that the Friesian had beef potential and the offspring were always fed on a restricted diet and by 2–3 years of age they had large skeletons (bone) and little flesh (meat) thus being very poor beef animals. Modern systems allow access to generally unlimited quantities of food (**ad lib feeding**) and allow the animals to grow rapidly to beef animals (much meat, less bone).

There are still distinct beef herds being kept in this country, with farmers rearing cattle in many different ways. However, the following description applies, in general terms, to a large proportion of cattle production in this country. Here it is implied that the dairy and beef production are parts of the same enterprise on a farm. This may not be so, for the farmer may sell the calves (or young animals) and another farmer will complete the beef production; however, the essential elements are the same.

In general, cows are able to produce calves at any time of the year. Domestication has brought with it the cessation of seasonality of reproduction that is still characteristic of sheep. A dairy farmer may, if he wishes, have cows producing calves regularly throughout the year in an attempt to maintain a regular milk yield. Milk is generally sold through the **Milk Marketing Board** and the farmer receives payment for this every month. This regularity of payment is unlike most agricultural enterprises and has provided a secure first rung for a number of new farmers. The price received for milk by the farmers is usually highest during the winter months (when production costs are highest) and so the farmer may be tempted to organise more cows to calve in autumn to provide winter milk.

The **oestrus cycle** is about 21 days in cattle, and the **gestation period** about 9 months (nearly identical to humans) and usually one calf is born. Cows will have their first calf at about 2 to $2\frac{1}{2}$ years of age.

The initial milk (**colostrum**) will be given to the calf for the first few days of life. The calf will then be fed milk substitute (so that the milk from the cow can be sold) and also encouraged to eat dry food, e.g. hay, as soon as possible. This

Figure 5.4

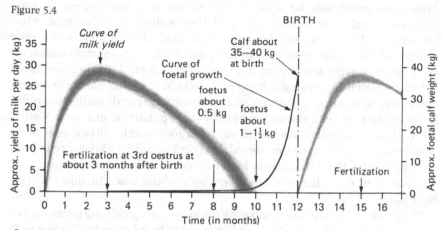

Curves of milk yield and foetal growth

latter process encourages the maturation of the rumen of the calf which is essential for eating grass. Milk substitute is made by reconstituting dried milk powder with water. It is an interesting facet of current milk price structure that it is cheaper to feed milk substitute than to use the cow's own milk to feed the calf. This arises because only about half the total milk of the UK is sold as liquid milk to the consumer. The other half is manufactured, being turned into butter, cheese, condensed milk, dried milk, cream and yoghurt; uses which command a much lower price to the farmer. (Price is not the only consideration for feeding dried milk to calves, there are other management factors which encourage maximum yield from the cow.)

The amount of milk that a cow yields rises daily to a peak some six to eight weeks after parturition. Generally this peak is held to about three months and then gradually declines until about ten months, when the cow will be 'dried off'. This leaves a period of about two months when she does not produce any milk until the next calf is born, for she will have been fertilised at about three months after parturition. Notice here that domestication has over-ridden any cessation of oestrus that may be caused by lactation which seems apparent in some other mammals.

The last two months, when the cow is not giving any milk, coincides with the maximum growth rate of the foetus, and so energy and protein which was being used to produce milk can now be used for maximum foetal growth, usually helped by extra feed. This is called steaming-up and helps to produce a good calf and maximum subsequent milk yield. These facts are summarised in Figure 5.4.

The shape of the milk yield during lactation could also be used roughly to summarise the trend of total milk yield per lactation throughout the cow's useful life. Generally, the first lactation has a slightly lower total yield than is subsequently achieved in the next two or three lactations. Then yield per lactation declines and eventually becomes uneconomic because the cost of food, labour and capital given to the cow is not recouped in sales of her milk.

These cows must then be replaced. The farmer may buy potentially good heifers from other herds and hope that they achieve their potential. Most farmers breed replacements from the herd itself. The farmer needs some estimate of the quality of his present cows and this is one important reason for recording **milk yield**. It is important to know the milk yield so that cows can be fed appropriately, for with declining milk yield in the latter part of lactation, feeding excess of expensive concentrates (see next page) could easily lose profits. Knowing the milk yield also allows the cows to be put into a rank order. Generally the best cows with excellent milk yield and good health will be used to breed replacements. A bull of the same breed (i.e. Friesian bull for a Friesian cow) will be used to serve these cows, or at least the semen will be used, because most fertilisation in dairy herds is nowadays by **artificial insemination** (AI) (see next page).

The less desirable cows (e.g. those with a lower milk yield) will generally not be used to breed replacements, but will be used to breed cross-bred calves with the aid of a beef breed (e.g. Hereford, Continentals).

This situation is summarised in Figure 5.5.

Feeding cattle

It is an idea of some antiquity that one should consider two aspects to the feeding of farm animals. There is the need to provide for basic physiological functions merely to keep the animal alive and maintain it in good health. On top of this, extra provision must be made for growth, milk production etc. This led to the idea of **maintenance ration** and **production ration**, i.e. certain mixtures of foods to keep the animal alive and other mixtures for producing muscle, milk etc. This distinction is not really true, though the idea that, for example, a higher yielding cow requires more food than a lower yielding cow is, in the main, true. It is not difficult to find agricultural literature that still refers to rations that provide 'maintenance +2', which means keeping the cow alive and also producing two gallons of milk, though the latter is usually

Figure 5.5

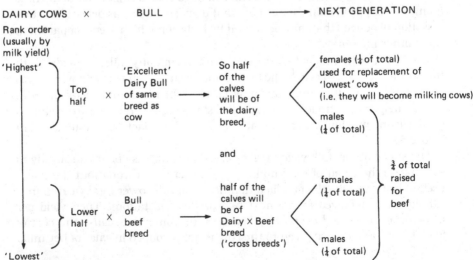

metricated to 10 kg (the density of milk is nearly equal to the density of water).

Foods for cattle can be classified as **roughages** and **concentrates**. Maintenance rations tend to have a high proportion of roughages; production rations a higher proportion of concentrates. Roughages include hay, silage, grass and root crops and are foods which contain a comparatively low amount of available energy and protein and a larger amount of indigestible material (equivalent to the fibre in human diets). Concentrates have a much higher proportion of available energy and protein, and are usually much more expensive. They include the cereals, as well as products from other food manufacturing processes, e.g. what used to be colloquially referred to as cake – oily and proteinaceous residues from the extraction of oil from seeds (cotton seed cake, linseed cake, etc.). These latter materials are not now generally available directly to farmers. Manufacturers use them to prepare foods (usually pelleted and in bags) of certain composition for feeding to different animals at different production stages (e.g. calf weaner nuts). These are a reminder that certain aspects of UK agriculture (and our lives) have been founded upon the exploitation of the colonies, some of the currently developing countries. Their crops, like the oil seeds mentioned, as well as sugar cane, spices and beverages, were purchased for low prices and shipped to the UK. In the UK they enabled the establishment and maintenance of, for example, a dairy industry in the hill-lands of the UK where inherent production is poor. These farmers were, in effect, able to 'buy in acres' at a very competitive price.

When feeding cows, the bulk of their diet in the summer consists of grass, and in the winter hay, or more usually silage available ad lib. Concentrates are provided individually for each cow at milking time. Precise quantities are given to match the milk yield, hence the need for accurate recording of milk yield. The latest rotary milking parlours are able to milk cows in a matter of minutes, and some farmers have expressed concern that the highest yielding cows will be unable to eat the concentrates they need in the short time that they are on the rotary milking parlour.

The animals which are being fed for **beef** may be treated in slightly different ways. For some, maybe those born in winter time, it is often advantageous to rear them **intensively**. They will usually have access to both concentrates and roughage from an early age, they will grow rapidly, probably remain inside the buildings and be ready for slaughter at about a year old when they weigh about 0.4/0.5 tonne live weight.

Others are reared **semi-intensively**, having an intermediate summer grazing between two winter periods. They will be about $1\frac{1}{2}$ to 2 years old at slaughter and weigh about 0.5 tonne.

Some farmers still produce beef **extensively**. These animals receive little more than grass or hay and often pass through a **store** period when they remain alive and healthy but put on little weight. They will probably be about three years old at slaughter and may not have the best carcase.

Artificial insemination (AI)
Since the Second World War, AI has increased rapidly as a valuable technique in the cattle (particularly dairy) industry throughout the world. Greater

physiological difficulties initially hampered its use in the pig industry, but it is now an available technique. The seasonal variation of ovulation in sheep also means that AI has no commercial advantage.

Bulls with excellent parentage are purchased by the Milk Marketing Board and kept at recognised centres. Here they regularly give semen which, after suitable treatment, can be deep frozen for long term storage. Once farmers notice signs of ovulation in the cows, a telephone call means that artificial insemination is possible within a matter of hours.

AI has allowed the rapid up-grading of the whole of the British dairy herd, as desirable bulls are now able to be used to inseminate many thousands of cows, rather than the few that would be traditionally available when one bull stayed with a herd of cows. It also means that cows are being inseminated with the semen of bulls that are long since dead.

Extensive cattle rearing

The intensive cattle rearing just described is typical of the developed countries. In the less developed countries this intensity is rarely apparent. In areas of **Asia**, the combination of high population numbers and the generally low standard of living means that fewer cattle are kept because of the inherent inefficiencies of cattle production.

In **Africa**, the cattle that exist are often kept by nomadic methods, particularly around the Sahara. The grazing is at best sparse, and often dependent upon seasonal and slight rainfall, so the people and the cattle have to move to the new grazings. This implies that the possessions of the people are minimal, essential and portable. In other areas nearer the tropics, the high rainfall allows the tsetse fly to survive, which transmits sleeping sickness (*Trypanosoma* spp.) to cattle and therefore effectively bars the area to cattle.

The indigenous cattle in these areas are usually slow growing, thin, with a poorly developed skeleton and little fat and muscle tissue. They are often kept as symbols of wealth, for their milk, sometimes for their blood, and at the end of their productive life for beef. They are better adapted to the hot climates of these areas and have a better resistance to parasites than breeds introduced from temperate areas. During the colonisation of Africa, some UK breeds (e.g. the Hereford) were introduced to improve both the quality and the rate of growth of these animals. In some areas, for example Zimbabwe, this has been reasonably successful and the cattle are kept by a system often known as **ranching**. Ranching, formerly characteristic of all American cattle raising, is still practised extensively in certain areas of the USA as well as in parts of South America, Australia, and South Africa. In these situations the natural vegetation is sufficient to provide nearly continuous grazing in an area, though often this is a large area and the line between nomadism and ranching is indeed fine.

Australia is an example of a country displaying a diversity of cattle keeping. Some areas are fenced and the cattle well managed. In other areas, the cattle roam over thousands of square kilometres, and are visited very infrequently, maybe only once a year when cattle are sorted for sale.

Cattle diseases

Whilst there are many diseases to which cattle are prone, **foot and mouth disease** is one of the most virulent and easily transmitted. It is caused by a virus which produces lesions at the mouth and feet. Whilst the disease is not usually fatal to adult cattle, it causes severe economic losses due to very reduced milk yield, loss of condition in beef animals and loss of calves. To prevent the spread of infection, an immediate slaughter policy is implemented in the UK. It also affects sheep, goats and pigs as well as wild herbivores.

Mastitis is an inflammation of the udder of the cow caused by bacterial infection. It is thought that inexpert milking contributes to the susceptibility of the cows to mastitis. At every milking, an initial quantity of milk is drawn from each quarter of the udder to look for any abnormalities which would be indicative of mastitis. Mastitis not only causes a loss of milk yield, but the milk is unsaleable and the cow must be milked separately; these are all considerable drawbacks. Farmers therefore practise a high standard of hygiene in milking parlours, and at the end of each milking the teats of the udder are usually dipped in an appropriate disinfectant (e.g. iodophor, hypochlorite solution).

A successful campaign to eradicate **tuberculosis** from cattle has helped to lower the incidence of the disease in humans in the UK. As the disease is easily transmitted between humans and cattle, its control is vital to human health, as is the pasteurisation of milk to kill pathogenic micro-organisms.

5.5 Pigs

Pigs are present, either in the domesticated form or the wild form (wild boar) in many countries. It is a valuable animal for several reasons:

(a) it will live in a wide variety of environments;

(b) it will eat a wide variety of food, even being a scavenger;

(c) potentially, it is quite prolific and grows to a useable size in months rather than years. However, in many areas it roams the huts and farmsteads, receiving little attention, and accordingly it is far less valuable as a food than in the developed countries of the world.

Whilst it is virtually absent from Islamic countries for religious reasons, approximately one third of the world's pigs are kept in China where they provide not only a valuable meat, but also a useful export commodity – bristles for brushes.

The **pig** is a **non-ruminant**, with a digestive system similar to man, though there is much greater bacterial digestion of cellulose.

Pigs are kept to provide **meat**. The fresh meat is called pork, and unless cooked (or preserved) will rapidly deteriorate. Pig meat has been traditionally preserved by curing with common salt, and this product is usually called bacon. Whilst the advent of refrigeration makes this method of preservation unnecessary, the distinctive taste of fried bacon ensures its continuance. Bacon which is cured by a slightly different process and subsequently boiled, provides the cold meat usually called ham or gammon.

Previously preference seems to have been, in general, for smaller joints of fresh meat, and larger joints for bacon (this term will also include ham). To meet these different demands, different breeds of pig arose in the UK. These were as a result of domestication and artificial selection from the indigenous wild pig, together with crossings with imported breeds. Some breeds, e.g. Berkshire, Middle White, were reared in conditions under which they matured earlier and gave smaller joints of meat, and were thus used for pork. Others, e.g. Large White, were reared in conditions under which they grew more slowly, matured later and so were used for bacon (legs for large, well-fatted hams; sides for large, fatty bacon rashers).

During the last few decades both the size of pork/bacon joints and the quantity of fat preferred by the consumers has diminished. Similarly, dislike of 'seedy', that is black spotted, bacon has led to the decline of pigs with black pigmentation, e.g. Berkshire which is completely black, Wessex Saddleback (belted) which is partially black with a white band or belt. The epithelial origin of mammary glands (even rudimentary ones, in the males) means that melanocytes are present in bacon cuts from the ventral regions of black pigs.

Modern improved conditions of feeding and management, together with considerable interbreeding, has tended to blur inter-breed differences. It is interesting to speculate on how much genetic difference there was (or still is) between the traditional breeds. Many points of difference between the breeds are only superficial (e.g. nose shape, coat colour, ear posture, tail curliness, etc.) and seem only to be perpetuated by competitions such as agricultural shows. All the evidence is that they have little correlation with meat quality. Many pigs in the past were kept in small numbers and fed on meagre rations. Under these circumstances, virtually any pig would only grow slowly and in time produce large fatty joints. These pigs would almost invariably become bacon. Any genetic potential for rapid growth/small joints that may have been present was never exploited because of the shortage of food. It is interesting that many pigs in this country currently being produced for pork (rapid growth/small joints) have the Large White breed in their ancestry, traditionally a bacon animal.

Whilst the nucleus of the traditional breeds still exists to provide some of the parental stock, the overwhelming majority of pigs now being used as parents of animals for meat are **hybrids**. The genetic variability present in the UK pig population means that many sows produce piglets with slightly dissimilar growth characteristics (the runt of the litter still occurs). This means that, from the same litter, under present intensive management systems, it is possible that the quickest growing and maturing piglets will probably go for pork, whilst the later maturing ones will go for bacon. Some farmers produce a larger, often fattier animal – the heavy hog, or cutter, the carcase of which is cut up for different purposes, e.g. pies, sausages, fresh meat, cured meat (bacon/ham), lard. However, it is always possible to restructure and tenderise the meat from any pig into attractive manufactured products such as sausages, pies, burgers.

Pig production

Whilst the wild sow probably produced one litter a year of about three or four piglets, the modern sow will produce at least two litters per year, each of eleven or more piglets. The sow has a gestation period of about 115 days – usually remembered as '3 months, 3 weeks and 3 days'. A **gilt** (young female that has not yet had a litter) will usually produce her first litter at about one year of age. Her useful reproductive life will extend to some three to four years (six to seven litters), after which smaller litters make her uneconomic.

Management skills after the birth (**farrowing**) and during the neonatal period seem particularly important in this farm animal. Piglets will often need to have their canine teeth (**tusks**) clipped, to prevent damage to each other and to the sow when suckling; to be given an injection of an iron compound to counteract anaemia; and to have their tails clipped to prevent subsequent tail biting. The sows are prone to lie down abruptly with little regard for their offspring, and frequently some form of crate (to restrict the sow) or low, horizontal bar is necessary to give protection to the piglets. Even then, sows frequently crush one or more of their piglets. Additional warmth is necessary for the piglets – usually an overhead infra-red light – which additionally attracts the piglets safely away from the descending sow. Whilst cows and ewes may need some assistance at the birth, they do not need the level of neonatal management necessary in pigs. This indicates some of the problems that would beset modern pigs if returned to the wild.

Whilst suckling, the piglets are also encouraged to eat dry food. This food is usually specially mixed and then pelleted. These pellets will be offered to the piglets at all times, so that they may eat as much as they wish (ad lib feeding). After a few weeks (maybe as little as three) the rapidly growing young pigs will be weaned. The sow will be removed from the pigs, and oestrus will recur within a week. After fertilisation, she will then produce her next litter within another four months. The growing pigs will continue to receive ad lib food, reaching pork weight (about 50–60 kg/1–1½ cwt) in about four months, or bacon weight (about 90 kg/2 cwt) in about six months. Both male and female pigs will be produced for meat. Usually the male pigs will be castrated (at about one to two weeks old) and are then sometimes called **barrows** or **hogs**. It is said that whole or entire male pigs (**boars**) give a distinctive taint (boar taint) to their meat when slaughtered. It is to prevent this, and possibly for reasons of suspected increased docility, that male animals are castrated. However, there is considerable evidence that entire animals gain more weight per kilogram of food eaten compared with castrated animals, that when slaughtered at a young age (up to about six months) boar taint is imperceptible and similarly that the animals are too young to display the adult boar aggressiveness. The problem is mainly of consumer education.

Pig systems

Like all farm animal systems, there are probably as many ways of keeping pigs as there are pig farmers. The simplest systems are the **free-range** ones where the pigs are free to roam the fields and have small round-roofed huts for shelter. Many pigs are housed, for it is assumed that protection from the wind and rain allows the pigs to continue to grow rapidly, rather than convert food into extra heat energy to maintain their body temperature under these inclement conditions. As food costs are the major expenditure (often 75–80%) in pig production, housing may seem sensible, and certainly it eases some management problems, as the pigs are easily accessible. However, close housing may facilitate rapid disease spread, and hot humid conditions are especially conducive to respiratory infections.

Waste disposal is also a problem in housed systems that is not present in the field system, and certainly some mechanisation is imperative in large pig units for what is otherwise a less than desirable manual job. Some systems are extensively mechanised, with the pigs housed on slats so that the urine and faeces pass through the slats and are then mechanically removed from underneath into adjacent tankers for subsequent spreading on the fields.

There is no clear evidence that any one system is any better than any other system. It is obvious that farrowing and neonatal management are important, as is disease control. These are some of the factors ensuring that the maximum numbers of pigs are reared. However, profit or loss may depend on other factors, e.g. food wastage (which may be under the farmer's control) or market prices and overseas competition (to which all producers are susceptible, with little personal control).

5.6 Poultry

There are two general characteristics of poultry:
  (a) they belong to the Class Aves (birds);
  (b) they are raised by farmers (wild birds are not included in poultry).

They are kept for their **eggs** and **meat**. Their **feathers** also have some uses, e.g. human bedding (eider down) or ornamentation. Poultry includes chickens, turkeys, ducks, geese, guinea fowl and pigeons.

Whilst birds have no teeth, their powerful **beaks** are able to gather and slice food, though pulverising (grinding) of, for example, whole cereal grains, has to be performed by a specialised part of the gut called the **gizzard**. The gizzard has thick muscular walls, is internally lined with hard tissue and also contains grit. The grit helps to grind the food, and would normally be picked up in wandering across the land. Birds in modern housed systems have no natural access to grit. There are two alternatives. The grit can be supplied regularly by the farmer (often as a soluble limestone which also provides some calcium for shell and bone growth). Usually, however, the food is mechanically ground and supplied to the birds in a powdered form, often called mash, so that the birds do not need to grind it themselves.

There are considerable numbers of the different types of poultry worldwide, with regional preferences (e.g. ducks in China). However, the **chicken** is numerically the most popular. The improved management and marketing techniques which have characterised the chicken industry during the last three decades are now being used to try to alter the traditional seasonal demand for turkey into an all-the-year-round demand. From earliest domestications, the chicken has existed closely with man and his dwellings, and still exists as such in many villages throughout the world. In these villages, it forages for most of its food: little is provided by man. The genetic potential of these indigenous breeds is poor, and with little food their growth is slow and the number and size of eggs is small, but this way of keeping chickens is extremely cheap. This is in contrast to the intensive systems which offer unlimited quantities of food and a protected, often controlled environment to chickens of high genetic potential. Here the chickens will grow rapidly and can reach marketable weight for meat at two to three months of age. Alternatively they will lay regularly (approximately daily) producing over 200 eggs from about five to six months of age.

Chicken is the usual name given to this species (*Gallus domesticus*), though **domestic fowl** has had previous popularity. **Chick** is used for the young bird (the use of chicken is not uncommon, though confusing). The **cock** is the adult male bird, and the **hen** the adult female.

Grower is a term sometimes used for the young growing animal, and may be confined to the females. Certainly, for the young growing meat bird, the term **broiler** is in more common usage. Broil means to cook over a fire or grill. This implies quite a high temperature and a quick process. This is usually called roasting and can only be used on young, and therefore tender, meat. So broilers are rapidly grown chickens (of either sex) for meat production, though they are usually marketed as 'chicken', at about 1½ to 2 kg.

Older birds fall loosely into two categories:

(a) There is still a small, somewhat seasonal and specialised (e.g. hotels) demand for a larger meat chicken (3–10 kg dressed weight). This demand is usually supplied by male birds. These are often castrated, normally by implantation of a pellet of female hormone in the neck skin. This **feminisation** encourages fat deposition, better carcase quality and a more subdued bird. This type of bird is also sufficiently tender to roast, and is sometimes called a capon.

(b) The female birds (hens) will normally be kept for the eggs they produce. After their productive life, which may be at only 18 months of age, they will be used for meat. As they are from egg-laying breeds, their carcase quality is not as high as that of the meat breeds. Some are sold as whole birds which must be cooked slowly and at a low temperature for the meat to become tender. This is usually achieved by boiling in water – hence boiling fowls; not to be confused with broilers. Most of these hens will be processed into chicken pies, meat paste, burgers, etc.

The modern techniques mentioned previously were initiated in the USA during the Second World War under the general impetus of a shortage of meat from other farm animals. Certainly prior to the Second World War there were no separate meat and egg production industries in the UK comparable with today's situation. The hens provided eggs and then meat; the male birds were fattened (often as capons) for seasonal meat production (e.g. Christmas). Total production of eggs and meat was much less than today, partly due to poor genetic potential (the animals had to be general purpose to provide both meat and eggs), and partly due to poor and inadequate nutrition. Chickens were usually merely a sideline, often of the farmer's wife.

Modern methods employ different breeds for egg production (light boned birds, with minimal muscle tissue; large egg production) and for meat production (rapid growth, particularly of muscle tissue, good food conversion ratio). These specialised breeds are usually **hybrids**. These exploit hybrid vigour so that the offspring display better characteristics than either of the parents. Also the farmer has to return to the original breeder to buy new stock, as breeding from hybrids does not produce similar offspring. Large companies are often both breeders of stock and producers of eggs and meat.

Most **egg production** takes place in buildings where the length of day is controlled, and often the temperature (and even the humidity) is regulated. An adequate temperature (about 20 °C) is essential if the birds are not to convert too much food into maintaining their body temperature rather than into egg-production. Day-length control is very important for modern egg production. The wild, undomesticated fowl (like many other birds) would usually lay a clutch of a few eggs in spring, and maybe further small clutches later in the year. The initial clutch in spring is in response to increasing day-length in temperate climates. The modern bird will come into lay at about five to six months (irrespective of time of year) if the day-length conditions are controlled. Whilst there are variations, most day-length control patterns have an initial continuous light (and warmth) period for the young chicks, decreasing gradually to about six hours per day (equivalent to mid-winter) at about four months of age, then increasing to at least 17 hours per day (equivalent to spring and summer). Under these conditions the modern hybrid hen, with appropriate nutrition, will produce well over 200 eggs in a laying season (compare this with the wild bird's clutch of a few eggs). The eggs will be laid approximately daily, but there are often one day gaps every few days. Egg production eventually ceases. Naturally, this might coincide with decreasing day-length in autumn. After this cessation of laying, the birds will **moult** (lose their feathers). This is a period of several weeks during which the hen has to be fed, but when she does not produce any eggs. Whilst she will subsequently grow new feathers and recommence laying, the numbers of eggs in the next laying season will be less than in the first, and this decline will continue until old age when laying ceases entirely. Most farmers regard even one moult as an uneconomical period as food represents the largest cost in egg production, and so hens will go for slaughter at the end of this first laying season.

Hens are usually kept in cages. These are usually in ranks, one upon another, and referred to as **battery cages**. There are usually several hens per cage. The cages are made of wire which allows the excreta to fall through and be removed by a scraper or moving belt. The floor of the cage slopes forward. Eggs which are laid thus roll forward out of the cage which keeps them away from the beaks of the hens, and allows easier collection. Food is provided just in front of the cage so that the hens can reach the food by putting their heads through the gaps in the wires. Water is provided by a special small tap which releases water when the hen puts its beak to it. Housing in cages represents the most economical way of providing the optimum environment for the chicken to produce the maximum number of eggs. This production of a high number of eggs must indicate optimum or near optimum physiological activity of the hen, which is not adversely affected by the caged environment. There is con-siderable evidence that farm animals in the UK are kept in better conditions than prevail in many other parts of the world.

The battery cage system just described is one example of an **intensive system**. Intensive systems are characterised by a high level of inputs – a phrase implying high food usage, high capital costs (houses, cages etc.) maybe high labour and running costs (e.g. electricity) on a small amount of land.

Another intensive poultry system is the **deep-litter system** – usually for broiler production (broiler houses) but used by some for egg production. Again the chickens are housed but this time they are free to roam throughout the house, walking upon absorbent litter. This litter can be wood shavings, peat, waste paper – anything which will absorb the chicken excreta and rot down (the fine particles present in sawdust irritate nasal linings and so should be avoided). Food is available ad lib, as is water. Again the temperature and humidity will probably be controlled. Faster growth appears to occur with increased day-length (some even advocating continuous lighting) which may just mean that the birds can see to eat more food in a given period, and therefore grow faster. Nevertheless, any small improvement in such a competitive industry may be economically important.

Factory farming

The intensive systems, particularly battery cages just described, are probably the most well-known examples of **factory farming**. Undoubtedly they represent a change from the previously-used methods of keeping poultry. As was pointed out, however, in all these systems, growth rate and egg production are of a very high order and animals would not thrive unless provided with the correct conditions. In the free-range systems the hens are exposed to all the different climatic conditions, and, like other birds, find it difficult to survive in the cold winter conditions. The health aspect of many free-range systems is not usually of the highest order, for often the chickens are kept as cheaply as possible. Whilst it is perfectly feasible to practise cleanliness, they are rarely treated for parasites because of the cost, and many wooden hen-houses harbour innumerable external parasites. These conditions are less likely to arise in battery cages because the parasites find few places to hide in metal cages and hygiene has to be of a high order, as any disease would spread rapidly through the entire flock because of their closeness. The large numbers of eggs produced from battery cages means that collection is more frequent and so, with the generally better standards of hygiene, battery eggs are more likely to be fresher and disease-free than those from free-range systems.

Foxes are unable to enter battery cages, though they still kill free-range chickens.

5.7 Sheep

Sheep provide humans with **meat**, **milk**, **wool** and **skins**. It is speculated that their flocking instinct was an important factor in their domestication by man, as it meant that man could more easily control them. However, in general, sheep in temperate areas still retain the annual reproductive cycle characteristic of undomesticated animals. In late summer/early autumn the ewes respond to decreasing day-length and ovulate. A pregnancy of about five months then produces offspring in spring to coincide with increasing food supply. The Dorset Horn breed are exceptional and are able to breed at any time of the year. In equatorial areas there is also annual reproduction. However, this is not related to day-length changes. Some workers have suggested that minor changes in the mineral content of the herbage during the dry season may trigger ovulation in some way. Certainly, lambs seem to be born to coincide with the vegetation produced by the rainy season. Sheep in general differ from cattle and pigs which will produce offspring at any time of the year.

The usefulness of sheep to man is indicated by their worldwide distribution, with probably over a thousand breeds (varieties). Goats, a near relative of the sheep, are also similarly utilised in many developing countries. In the UK there are over 30 breeds of sheep together with a number of cross-breeds. This diversity within the UK is indicative of both the variety of environmental conditions present, and of the pioneering work of sheep improvement that has been part of UK agriculture for several centuries and has supplied animals for newly-colonised areas. For example, the Romney has been successfully used as the basis for the New Zealand sheep industry as it seems to thrive well in humid conditions.

The Merino, a Mediterranean breed, provides the basis of the Australian fine wool sheep industry because of the similar environmental conditions. In the USA, some of the larger breeds, e.g. the Suffolk, have been popular because of their large size and good carcase quality. These breeds are therefore the result of both natural and artificial selection. This wide diversity prevents a complete description here of all the UK breeds, and standard agricultural textbooks should be consulted for further details.

Whilst any simplification is prone to error, it is possible to arrange the UK sheep into two groups related to their ecology: **mountain sheep** and **lowland sheep**.

Mountain sheep

The mountain sheep (e.g. Scotch Blackface, Welsh, Swaledale) are generally small (e.g. a mature ewe may weigh about 35 kg) and very hardy. Hardiness exhibits itself in the form of a dense fleece offering excellent protection against rain, wind and snow, and the ability to survive on sparse vegetation. It is this latter point which highlights the importance of these mountain breeds – they are able to graze where no other animal (of agricultural importance) could survive, and certainly where cultivation of the land is either physically or economically impossible. They are therefore providing food for humans where there is no alternative. The only other possible use for this land in the UK is afforestation. It must be realised that worldwide there are also many other places where the grazing of animals is the only feasible agricultural activity, e.g. savanna/sahel areas. Attempts to grow crops that could be consumed by humans are rarely successful. So arguments which indicate the loss of energy in converting herbage into animal tissue can only be acceptable when there are these alternatives, not when it is grazing or nothing.

One can speculate that these mountain breeds are very similar to their ancestors of many centuries ago. Each flock remains on its own territory without fences, and this ability must be passed on to subsequent generations. Certainly attempts to introduce sheep from other areas would be fraught with tremendous difficulties, and the sale of a mountain farm implies that the sheep will remain and be bought with it.

These mountain sheep are generally slow-maturing and rarely produce meat of the highest quality. However, they provide a nucleus for the production of cross-bred animals which, in less extreme environments, are much more likely to produce meat of desirable quality. On the mountains, the farmer hopes that each ewe will produce one lamb every year in spring. This would be a lambing percentage of 100. This is rarely achieved, 80% in good years, down to about 40% in the poorest years are the usual values. Twins would certainly present too much of a burden for the ewes under these arduous mountain conditions. The ewes will produce their first lamb when they are two years old, and after three (possibly four) breeding seasons they are **drafted** (selected from the flock). These draft ewes are sold to farms with a more favourable environment. Here the better conditions allow the ewes to produce lambs for a few more years. Also, their fertility increases (as a group, they have a higher lambing percentage, with some producing twins) and their milk yield is higher. Rams of larger, faster growing breeds (lowland) are used on these draft ewes, and the resulting offspring will often have a high growth rate for meat production, and some of the female offspring may themselves be used for breeding – possibly under the even better environmental conditions of arable farms, e.g. in eastern and southern England.

Whilst the mountain grazing can support all the sheep during the summer months, it can only support the adult sheep during the winter. All the lambs born in spring must either be sold in autumn (usually to lowland farms where the better conditions allow them to be fattened for meat) or some selected females, to replace draft ewes, will be overwintered on better lowland pasture. If this is not available locally, then these sheep may have to be transported many kilometres to appropriate quarters.

The mountain sheep are sometimes described as being at 'the top of the pyramid', for they are both physically on the mountain tops, and their increased numbers then move down to provide the basis of many lower flocks.

Lowland sheep

The distinction between mountain and lowland breeds is certainly not clear (there are several hill breeds, and cross-breeding with draft mountain ewes blurs the differences). However, others would argue that there are such considerable differences between these lowland breeds that it is incorrect to discuss them together. However, they all share a few characteristics. They will be kept on farms enclosed within some form of hedge, fence or wall, unlike the mountain sheep. This is a sign of a different approach to the management of these animals. They have to fit into the farming system alongside other crops and animals. Unlike the mountain sheep, which were the only feasible agricultural activity, these lowland sheep exist in competition with other agricultural activities. Whilst the breeds and cross-breeds are all different, and it is probably true to say that there is one to fit any farm or farmer, their economic viability usually means that they are capable of rapid growth and of producing quality meat (or transmitting these characteristics to their offspring).

Generally, they are bigger animals than the mountain sheep (some will be twice the weight at approximately 70 kg). The value of their wool is negligible (because of its low quality) compared with the value of the meat. Quality wool is more likely to come from other breeds in other countries (e.g. Merino: Australia).

With such a diversity of breeds and environments, it is difficult to generalise about keeping lowland sheep, but the following represents the general pattern.

Like the mountain sheep, the ewes will ovulate and be mated in autumn, so as to produce lambs in spring. Under these more favourable conditions (less cold winters, more food of better quality) farmers will be hoping for lambing percentages approaching 200 (i.e. the majority with twins) though most will achieve 160–190%. The feeding of better quality food for a few weeks prior to ovulation raises the number of twins. This is called **flushing**. Many farmers now house their ewes for the two to three months prior to lambing. Whilst this housing may be quite simple (covered roof, straw bale walls) it affords protection to the ewes during the later part of the gestation period when the foetus is growing rapidly, and management and feeding are much easier.

After lambing, provided the weather is mild, the ewe and lambs will have access to pastures. Ewes with a good milk yield will enable their lambs to grow rapidly. Gradually the lambs will eat more and more of the newly-grown grass. The lambs may graze ahead of the ewes, having access through narrow spaces which the ewes cannot pass. This allows the lambs to have the young fresh grass, and they are also less likely to pick up the intermediate stages of parasitic gastro-intestinal worms.

With some breeds in some areas the best lambs will have reached an acceptable weight and quality to be slaughtered for meat at about 12 weeks of age. This will be at about the end of June/beginning of July, and they will weigh about 35 kg (approximately the same weight as a mature mountain sheep). Many, however, will not reach an acceptable standard until the summer or autumn months. Others, under even poorer circumstances or with a slower growth rate, will have to be fed on various root/leaf crops (e.g. turnips, kale, sugar beet tops) before they are marketable during the winter months. This was the traditional type of large lamb that was produced by the Norfolk four-course rotation. The roots (of the roots–barley–seeds–wheat rotation) provided the fodder for these animals. This type of large, probably tough-meated, animal is not preferred today. Certainly this type of animal will be of a much heavier weight, probably over 50 kg but there is always the risk that the animal will have too much fat and bone and insufficient muscle (meat).

Sheep diseases

Mention was made previously of attempts to prevent lambs being infected with parasitic gastro-intestinal worms. There are several species of small **parasitic roundworms** which are usually present in all sheep in the fourth stomach and small intestine. When the infestation becomes heavy the sheep will lose condition, and have diarrhoea, a harsh fleece and poor body development. The adults in the gut produce eggs which are passed out to the ground in the faeces. In a few days, under favourable conditions, these eggs develop into larvae which eventually ascend the herbage and are ingested by other sheep to complete the life-cycle. Many larvae are thought to die within a short time if not ingested, and it is for this reason that lambs are sometimes allowed to feed alone (p. 62) on pasture which has not been recently grazed.

It is virtually impossible to guarantee breaking the life-cycle in this way and thus preventing both new and further infestations of animals. However, chemicals have been developed which can be given to the sheep to kill these internal parasitic worms. They are often given in liquid form via the mouth. Sometimes the name **anthelmintics** is given to these chemicals, for **helminths** are parasitic worms. Similarly **nematicides** is often used because some of these parasitic worms belong to Class *Nematoda*.

By a combination of good husbandry and chemical dosing these diseases can usually be contained.

A similar integration of methods also applies to other diseases to which sheep are prone. In **liver fluke** disease the adult flukes live in the bile duct of the sheep. Here they are difficult to kill and so measures to eliminate the intermediate host (*Limnea truncatula*) from wet pastures must be practised.

Prevention of some diseases has been comparatively successful. For example, **pulpy kidney disease** can affect lambs at an early age, and is caused by an infection of *Clostridium welchii*. However, vaccination of the ewe before birth provides immediate passive immunity for lambs in the colostrum of the ewes, and is then supplemented by vaccination of the lambs themselves.

Sheep scab is a particularly nasty disease of rapid onset caused by mites belonging to the genus *Psoroptes*. These puncture the skin and feed on the emerging serum. These wounds become inflamed and intense irritation causes the sheep continually to scratch and rub itself against any available object. Secondary infections of bacteria occur and the disease can then be rapidly debilitating. During the last few decades regular dipping (complete immersion) of sheep in a solution of **acaricide** (usually BHC) eliminated the disease from the UK and dipping was abandoned. However, the disease has reappeared and regular dipping has been reintroduced.

5.8 Fish

Even in the hunter-gatherer era, man caught and ate **fish**. This still continues today, though in many different forms compared with the early hunting of fish.

Fishing is a word used to describe all ways of catching fish, from rod and line and spear, to large deep sea fishing vessels which catch fish that man never tends. The first and only direct relationship man has with these fish is when he catches them (see *Man and the environment* in this series). This provides the bulk of all fish eaten by man.

Aquaculture or **fish culture** is the name given to the tending of marine and freshwater animals that are subsequently eaten by man. It implies some degree of husbandry, often the provision of shelter and food. This has an ancient history for the Romans cultured oysters, and it is possible that the Chinese were rearing carp in ponds as far back as the 5th century BC. On a small scale this has continued through the ages, for example the carp ponds in medieval monasteries.

Aquaculture involves different organisms in different parts of the world. Part of the reason for the differences is due to the different environments (mainly temperature), but consumer preference is also an important point. In the UK, carp were cultured in ponds and the moats of monasteries and other large houses some centuries ago but the taste for carp does not now exist to any measurable extent, for despite their excellent taste, their many fine bones are unacceptable. In all aquaculture, economics plays a very important part. This is particularly true in the UK, where, as an island surrounded by sea, the people have been used to the ready and regular availability of marine fish, e.g. cod and plaice. Whilst research work has shown that some of the marine fish can be cultured, often the cost of growing the fish is greater than the cost of fish landed at our ports. So the only fish that are cultured at present are the more luxurious ones like trout and salmon. Salmon farming in sea-water pens in Scottish waters is slowly becoming established, but unpredictable maturity has been a problem. **Trout farming** using rainbow trout has become established in the UK for the production of what is essentially a luxury fish even though its price is often very competitive with some of the more exotic marine fish such as sole and turbot. However, the future of trout farming is precarious. It currently supplies only about 1% of the fish consumed in the UK, and it needs high protein food to produce it. The food for trout contains nearly 50% protein which makes it expensive even though this protein is mainly composed of poor quality marine fish not wanted for human consumption, for 'Essentially, trout farming is a means of converting trash fish (the food) into high quality fish' (Sumpter and Woods, 1981).

Apart from the cost of the food, there are other problems to be solved for successful trout farming. An ample and clean supply of fresh water is essential and its temperature is critical, for growth at 18 °C is about double that at 10 °C. This can make economic success or disaster. The fish do not spawn naturally and both males and females must be 'milked' to effect fertilisation. The article previously quoted should be consulted for fuller details of trout farming.

Oyster culture has a long history, for they are regarded as a staple seafood in Japan, whilst they are a gourmet delicacy in the UK. Culturing, rather than mere collection, has evolved because better, regularly shaped shells are available from culture compared with the wild shells, and a better yield is possible with the advent of **vertical culturing** techniques. Fertilised eggs hatch into minute free-swimming larvae which eventually settle and remain sessile throughout the remainder of their life. If sticks, or any form of suspended collector are used, the larvae will settle on them, offering both a greater surface area within a given area, and also raising the oysters from the benthic predators.

Traditionally, *Ostrea edulis*, the European flat oyster, is cultured in the UK, but attempts to introduce the Pacific oyster *Crassostrea gigas*, which is faster growing, seem to have failed as it is sufficiently different from *O. edulis* to be discriminated against by the gourmet, though *Crassostrea* spp. have a firm hold in European markets. Being a **filter feeder**, water pollution may present a problem to oyster culturing. However, it is usually only acute pollution which is the problem, and sometimes low level pollution may be beneficial if it encourages algal bloom and therefore more food for the oysters.

An interesting example of a marine organism that can be cultured in fresh or brackish water is the milkfish (*Chanos chanos*) which is produced on a large scale mainly in Indonesia, the Philippines and Taiwan. Adults do not spawn in captivity, and so fry have to be captured from coastal and estuarine waters. They are euryhaline and can be successfully reared on algae in freshwater ponds.

Despite its lack of acceptance in some countries, carp (several species) is a popular fish in many parts of the world, particularly Asia. It spawns relatively easily in captivity, is hardy and fecund and grows rapidly. It can be fed as a herbivore, and therefore much more cheaply than, for example, trout. As it is omnivorous, it readily eats small animals and therefore is beneficial in removing, for example, disease vectors like mosquito larvae. It is usually cultured in ponds, but many exist in paddy fields and provide an important protein supplement to the usual staple diet of rice.

Over a dozen species of *Tilapia* are regularly cultivated throughout the world. In recent years they have grown in popularity in such places as S.E. Asia, Japan, India, near East, Africa and parts of Europe, USA and South America.

They are tropical freshwater fish, some tolerating brackish water and some tolerating the cooler water of non-tropical areas. Belonging to the family Cichlidae, many *Tilapia* have created interest because they are mouth-brooders. Their spread in popularity is undoubtedly due to their ease of culture as little human skill is needed. They spawn very easily, and often overpopulation is a problem. This produces many stunted fish, which may not find a ready market, particularly in Africa.

They are mainly herbivorous, with some being omnivorous. Several species will survive in waters which have such a high organic matter content that other edible fish would not survive. Supplementation of the water with inorganic fertilizers e.g. basic slag, superphosphate, and any form of organic wastes such as rotten fruit, mill sweepings, kitchen waste, produces significant increases in yield. Sewage will similarly increase yield of *Tilapia*, but only seems an acceptable practice in South East Asia.

6 Land reclamation

6.1 Introduction

The continued increase in population poses two interrelated but conflicting problems. There is the need to provide more physical facilities (essentially roads and buildings) for people. This consumes land which, in the main, has been used for agriculture. There is also the need to provide more food for this increasing population. So the total food requirements are increasing and this food is being produced on a diminishing area of land.

Attempts to increase the amount of land available for agriculture are often referred to as **land reclamation**. This term is not consistently used with precision, and in some cases this refers to an initial claiming, for example, land claimed from the sea which has never previously been used for agriculture.

Reclamation of land that has been used for other purposes and is then restored to agricultural use usually occurs after mining or quarrying activities. Whilst many gravel workings are restored after extraction, this is often to an amenity value, rather than for agriculture. However, some clay pits are being restored to agricultural use (p. 68).

6.2 Mining

For the purposes of dealing with land reclamation, mining activities fall roughly into two categories:

(a) opencast coal mining; with some iron-ore opencast mining which present similar problems;

(b) older methods, involving both coal and non-ferrous metals, usually characterised by extensive spoil heaps.

Opencast coal mining is a comparatively recent technique, having begun in the UK during the Second World War. From the start, restoration to agricultural (or former) use was one of the priorities of opencast mining which had never been considered in the previous deep working/spoil heap mining.

Modern machinery is capable of removing large quantities of soil in fairly accurate layers, so prior to the opencast mining operations approximately 30 cm of top soil are removed and stored in heaps, which are often placed to act as noise barriers. For more effective restoration at a later date it is important that this soil is moved at a suitable time and in a suitable manner. If the soil is too wet, it is much more likely to compact and become structureless and anaerobic. In poorly planned operations machinery may operate continually on the same part of the heap which will similarly compact.

After the removal of the top soil, approximately 60 cm of sub-soil are removed and similarly stored in heaps for later replacement.

Extraction of the coal takes place in parallel cuts, with the **overburden** (unwanted layers) being moved to the area from which coal has been removed. Thus the bulk of the overburden is moved only once (see Figure 6.1).

After the removal of the coal and the replacement of all the overburden, the height of the site will be very similar to the former level. The loss of the coal

Figure 6.1

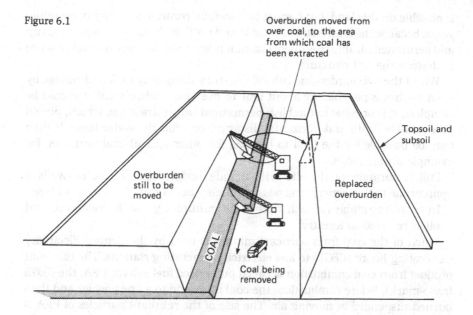

Overburden moved from over coal, to the area from which coal has been extracted

Topsoil and subsoil

Overburden still to be moved

Replaced overburden

COAL

Coal being removed

(possibly in the ratio of about 1:20 to 1:30 of coal to overburden) is compensated for by the increase in volume of the overburden brought about by the breaking up of its structure during movement. Settling will take some three to five years.

After levelling, the overburden is **scarified** (using a large deep-toothed rooter-cultivator) to break up any compaction (panning) and remove large rocks.

The subsoil is replaced (usually in two layers, each layer being scarified) to provide the 60 cm layer of subsoil. Ideally this will be a summer (dry weather) operation, and this subsoil is left rough after rooting, to overwinter and weather.

These subsoils are normally excessively acid, so lime (often ground limestone) is applied to counteract this acidity. The top soil will then be replaced in a 30 cm layer. This soil is also likely to be acid and lacking in nitrogen and phosphates, and thus needs lime and fertilizer. The fact that the soil has been in a heap often contributes significantly to its acidity and the low nitrogen and phosphorus levels. The mere fact of piling soil into heaps, even when done under ideal conditions, causes compaction which makes most of the heap anaerobic. Thus, instead of the normal aerobic microflora and microfauna which help maintain the soil fertility, anaerobic organisms, which characteristically produce acidic metabolic products (see silage making), thrive and render the heap acidic. Compaction also produces a structureless soil which it may take much cultivation to remedy. Where possible, subsoil and soil are relaid as soon as possible so that they are in a heap for the shortest possible time.

During the late spring/early summer cultivating continues and a grass/clover mixture will be sown. This will often be cut and conserved in subsequent years, and regular lime/fertilizer treatment will continue. Grazing is possible on this land, but it must be carefully controlled during these initial years, because the land is still not stable and until the humus content builds up and permanent drainage is undertaken, it is very easy for excessive animal use to destroy the soil structure.

Whilst the overburden and subsoil are still settling only surface drainage by open ditches is possible. By about four to five years, when settling should be complete, it is possible to establish permanent under-drainage, which, placed at a suitable depth and horizontal distance, controls the water table. It then may be possible for the land to be used for other agricultural purposes, for example arable crops.

Full restoration of the site will include hedges and fences, or walls if appropriate to the scenery, and possibly shelter belts or small spinneys or trees.

In areas unsuitable for such intense agriculture, e.g. South Wales, the land is often restored to forestry.

Much of the coal from opencast mining is used by the Central Electricity Generating Board (CEGB) to fuel electricity generating stations. The resultant product from coal combustion is called **pulverised fuel ash** (or PFA, the CEGB trade-mark). Before combustion, the coal is ground to a fine powder and then burned suspended in moving air. The size of the resultant particles of PFA is about half and half of fine sand (0.2 to 0.02 mm) and silt (0.02 to 0.002 mm). PFA thus has promise as an agricultural material because this mix is similar to that part of the soil which provides available water for plants.

The PFA is composed mainly of the soil in which the coal measure plants originally grew, and is similar in composition to many present-day soils.

This material not only finds use in the building and construction industries (e.g. in cement, in building blocks and road material) but also is being used to fill pits from which other materials have been extracted. For example, southeast of Peterborough the London Brick Company is using it to fill in old brick clay pits. Transported in railway tankers, the PFA is mixed with water on site and the resultant slurry pumped into the pits where it spreads, levels out and subsequently dries. Although the PFA is sterile and virtually inert, it is possible with suitable treatment with fertilizer, to grow crops directly in the ash, though usually some 7 to 10 cm of top soil are added for best results. The management techniques are then similar to those described for restored opencast workings. At Peterborough, the top soil comes from the nearby sugar beet processing factory, having been washed from the sugar beet.

The spoil heaps of older mine workings present different problems. Much of this mine working took place in areas which were less satisfactory agriculturally, and so may hold out little promise of useful return to agriculture. Often a complete levelling and landscaping would present both financial and topographical difficulties for many spoil heaps are surrounded by roads, buildings, canals etc., which would make the spreading of the material difficult. Often heaps are rounded-down and sown with grass for the best aesthetic compromise.

The spoil heaps from non-ferrous metal workings have additional problems, for the level of residual metal in the spoil heaps often inhibits plant growth. Even after spreading, where this is possible, and covering with top soil, the toxicity still inhibits deep root formation, and thus plant growth is poor. However, it has been possible to isolate tolerant varieties of grasses which have established themselves on the edges of these 'metallic' spoil heaps. These varieties can tolerate much increased levels of various metals and represent an interesting example of micro-evolution. It is postulated that plants adjacent to the heaps produced seeds with a variety of characteristics, and some of these produced plants which had a high metal tolerance. These plants, in true Darwinian fashion, were fitter on the spoil heaps and survived. Seeds from these tolerant species can be used to reseed this type of spoil material.

6.3 Sea and wetlands

Whilst reclamation of the type just mentioned is of very recent origin, reclamation of land from the sea or adjacent wet lands has much older origins. Techniques for the reclamation of land from the sea are very similar, though in general terms two different types of land have resulted, as exemplified by the Polders of the Netherlands, and the Fens of Britain.

The formation of the Polders began several centuries ago and one of the main aims then, as now, was to prevent the sea flooding adjacent land during periods of exceptionally high tides and winds. The coastline of the Netherlands was very irregular, and the expedient of building a defensive wall across the mouth of an inlet provided not only a shortened sea-wall to defend and maintain, but also an area now cut off from the sea. Walls erected to keep out water are often called **dykes**, though the word can also mean ditches dug to assist drainage.

The Polders that have resulted are usually totally enclosed by dykes so that neither sea water nor river water have direct access. This land is below high water level, and some below low water level, so provision for drainage is essential. This has been provided in the past by wind power, with windmills driving Archimedian screw pumps. Fossil fuels are now superceding wind power. Eventually land thus enclosed dries out sufficiently to allow agricultural activities, and the only source of water for the crops is rainwater, and water entering through the soil by gravitational and capillary infiltration from surrounding areas. The pumping operations can thus maintain the water table at a level suitable for agricultural use.

Some initial problems may be encountered with **salinity**, though this usually disappears as rainwater gradually leaches out the excess salt.

The Fenlands around the Wash in the UK were reclaimed by similar techniques. However, the initial stages and situation were different from those initially present in the Netherlands. The land was mainly marshes and above high water level. Water from inland entered these areas in rivers which meandered slowly to the sea. The cutting of wide and straight (and therefore shorter) new courses for these rivers, for example, Morton's Leam, Vermuden's Drain, enabled the drainage of these marshy lands and therefore allowed for a wider agricultural use than the previous intermittent grazing and cattle production.

Except for clayey outcrops where villages were established, the most productive land had a high peat basis. **Peat** is organic and rather porous and so drainage causes some shrinkage due to porosity, but the main loss has been due to its gradual oxidation. The peat was formed as the result of anaerobic conditions when the vegetation died. Drainage and cultivation allow aerobic conditions to predominate and so oxidation of the peat occurs which releases inorganic nutrients which make the soils very fertile, but it results in the disappearance of the peat. The classic case of a post at Holme Fen which was erected last century with its base on the underlying clay shows this loss of peat. When the post was erected the depth of soil was over 6 m. It is now only about 3 m. Hence the Fenland adage that the soil level drops at the rate of a man's height in his lifetime.

So, from a situation in which drainage was by gravity, the lowering of the land surface of the Fens now produces a situation similar to the Dutch Polders. Initially it was sufficient to provide sluices at various points along the rivers. These acted as one-way valves, allowing the river water out at low tide, and closing to prevent sea-water entering as the tide rose. There are now many areas where the level of the land is such that continuous pumping has had to be adopted. In many situations the river bed is above the land level and runs in embankments. Consequently water has to be pumped uphill into these rivers, or out to sea.

Some areas of the Wash adjacent to the sea have been reclaimed because favourable conditions have allowed solid material to accumulate. These are sheltered areas, with very little natural fall and so suspended material in the very slowly moving water tends to accumulate. Eventually beaches form upon which dykes can be built to enclose the land – a situation very similar to the Polders.

Whilst reclamation of the Fenlands is, in the main, not of recent origin, modern reclamation often produces a conflict of interests. In the Somerset levels, the extraction of peat is destroying the surface vegetation and related grassland ecosystems, and plans to improve the drainage of grazing pastures so that they can be cultivated has aroused opposition from those who regard the present ecosystem as unique and hence worth protection.

6.4   Irrigation

Irrigation is an ancient technique and large areas are producing food because of the diversion of river/stream water onto fields, e.g., paddy fields; the land alongside the Nile.

Thus irrigation can be regarded as a land reclamation technique for it allows the agricultural use of otherwise unproductive land.

The reclamation of arid desert areas has been suggested. Attempts to turn some of these areas into productive land cannot necessarily be achieved just by causing water to flow into them. These areas are often exposed to regularly high temperatures and low humidity, and thus the potential evaporation of water is high. If irrigation water is applied to level land with a relatively impermeable subsoil, or insufficient water is applied, then the flow of water will not be downwards through the soil. In these cases, water will be lost from the surface of the soil by evaporation and not by drainage. The incoming irrigation water always carries dissolved salts, and so, as the water evaporates, these salts will accumulate in the surface soil. Areas of India and Pakistan are now too saline for plant growth. Nevertheless, correctly used, irrigation has considerable potential for both bringing land into production and for increasing yields in already cultivated areas (Ambroggi, 1980).

In the UK, studies have shown that, particularly in the south-eastern areas during the summer months, there is considerable potential for irrigation. This is mainly as a supplement to rainfall, which, being unevenly distributed through the year, is usually least during the summer months. The successful use of irrigation depends upon the cost and availability of water, as well as the correct choice of crop to irrigate. Generally, the most promising crops are those whose desired yield is a vegetative part of the plant, for these contain a high proportion of water. For example, potatoes, carrots, onions, grass may show a response to irrigation which is economically attractive.

Crops where the yield is a low moisture content seed or fruit, e.g. cereals, usually show only a small response to additional water and irrigation is uneconomic.

6.5 Land from forests

Many people have assumed that the **tropics** provide an ideal environment for crop growth, because the lush natural vegetation of trees, bushes and shrubs grow so well in the ever present heat and frequent rainfall. Little consternation was therefore initially shown when increasingly large areas of the tropical forests were felled to provide timber, partly for consumption in the home country, but mainly for export. It was assumed that these areas would provide large areas for crop growth and thus an increase in the world food supply.

However, this lush natural vegetation is not necessarily a useful indicator of excellent crop growth. In the tropics, the vegetation represents the main part of both the **organic** (about 80%) and **nutrient** (about 60%) status, with only a small portion in the soil. When the vegetation is removed, the humus/nutrient status of the soil is only sufficient to support the crops for two or three years. The **humus** is rapidly oxidised under the continual high temperatures and moisture, and soil structure disappears allowing erosion to take place. Therefore successful agricultural utilisation of newly acquired land from **tropical forests** depends upon measures to maintain the humus content of the soil, e.g. use of composts, green manuring, and reducing erosion. **Erosion** can be reduced by various practices:

(a) continual plant cover (also helps organic matter status) because roots bind the soil and leaves cushion the rainfall;

(b) contour farming in which the land is ploughed around a hill at right angles to the slope, instead of up and down. Water settles in the furrows and then sinks through the soil rather than flowing rapidly down inclined furrows;

(c) terracing is another form of contour farming. Here wide cultivation strips or steps are built along the contours. Here again the water stops on the strips and seeps through rather than rushing downhill.

7 Possible conflicts

Some of the ways in which agriculture impinges upon society have been mentioned previously, for example, factory farming, drainage of wetlands, etc., but there are also other areas of conflict.

7.1   Hedgerows

Concern has been raised at the removal of many **hedgerows** in recent years. Whilst some hedges have origins of over 1000 years, many have only existed for about 200 years. They were planted when land was enclosed; hawthorn was mainly used because it grew quickly to give a stock-proof hedge. At that time there were many areas of **uncultivated land** where many plants and animals existed. One of the concerns now is that this reservoir of uncultivated land has virtually disappeared in many areas, so that the hedgerows are the only remaining areas of uncultivated land. These hedgerows are the habitat of a variety of plants as well as many different animals. Some of these represent the lower trophic levels of some food pyramids, supporting animals which control pests which damage crops in the field (e.g. owls are supported by hedge-row animals so that they are available to eat mice; birds are supported by a variety of hedgerow food so that they can eat caterpillars that attack crops). Whilst modern cereals are self-pollinating, other crops (e.g. field beans, and some fruit crops) need insects for pollination. These are also supported by the hedgerows. The counter argument is that hedgerows are reservoirs of weeds and pests which spread into the fields and compete with the crops. Certainly arable fields have always had a very much reduced number of species (ideally, from the farmer's point of view, only one), with weeding being practised from the first settled agriculture; by hand, then mechanically, and now by herbicides.

Problems are said to have occured after hedge removal because of the lack of wind-break effect. High winds over newly sown but otherwise bare lands have allowed dry soil to be moved by the wind. However, the effect of the hedge as a wind-break may extend only over a short distance; probably about ten times its height (i.e. about 10 m each side of a 1 m high hedge). Thus wind-blown soil still occurs even in some of the smallest fields and personal experience suggests that this will happen even in fields less than 100 m wide protected from the prevailing wind by coppices. It is possible that the removal of hedgerows to increase the efficiency of continuous cereal growing is coinciding with a slight decline in organic matter in the soil. Whereas rotations usually maintain the organic matter at a certain level, this level will decline with modern methods of cereal growing when little organic matter is returned to the soil. Instability of structure may develop when soil organic matter falls below 3%. Levels of 2–3% organic matter have been found in soils under cereal cultivation in which the only returns of fresh organic matter are from root systems and the residues of their tops. This may leave insufficient organic matter to maintain soil structure, and, with a breakdown in soil structure, the individual particles in light soils do not adhere to each other and thus are more likely to be blown away when there are no plants to protect the soil.

It is certainly difficult to know where the balance between advantage and disadvantage lies. Possibly a much better argument, because it considers a different aspect, is to say that hedgerows are aesthetic and so it is important to preserve them because we enjoy the variety of plants and animals found there.

7.2   Recreation

It was mentioned earlier that gravel pits are often restored as an amenity rather than for agricultural use. They provide a facility for water sports, fishing and pleasant walks on their banks. This is not really a land use conflict, but conflict does arise when the public wants access, for recreational reasons, to farming land. In general, ownership of land is in private hands and the public only have the right to pass along public footpaths, and in some areas it is only comparatively recent actions like the 'mass trespass' on Kinder Scout some 50 years ago which produced such rights. The passage over other parts of farmland that are not designated a public right of way is theoretically a trespass, but in general the only legal remedy a farmer has is for a costly, privately fought civil action in which he would have to prove damage. Thus the public wanders over areas which legally it should not, but urban dwellers feel that they have an ancestral right to do this for general relaxation from the stresses of life.

7.3   Farm chemicals

Many chemicals are now used on farms for a variety of purposes. All have contributed to an improvement in quality and the yield of agricultural produce which would have been impossible by other means. There are concerns, however, that these chemicals may have side-effects which are detrimental. Whilst there are notable examples, such as the effect of DDT on eggshell thickness, and Aldrin and Dieldrin killing predatory birds, there is little reliable evidence of deleterious effects of chemicals that are currently used (Broadbent, 1980).

Further Reading

Ambroggi, R. P. (1980) 'Water', *Scientific American* 243(3)

Bardach, J. E., Ryther, J. H. and McLarney, W. O. (1972) *Aquaculture*, Wiley

Bowman, C. (1977) *Animals for man*, Arnold

Broadbent, L. (1980) 'Ecological aspects of agricultural pesticide application', *Biologist* 27(3), 131–3

Cole, H. H. and Garnett, W. N. (Eds) (1980) *Animal agriculture* (2nd edn), Freeman

Finch, R. (Gen. Ed.) (1977) *Fertilizers*, ICI

Green, M. B. (1979) 'The energy balance of pesticide use', *Biologist* 26(3), 123–6

Greenland, D. J. (1984) 'Rice', *Biologist* 31(4), 219–225

Hartmann, H. T., Flocker, W. J. and Kofranek, A. M. (1981) *Plant science*, Prentice-Hall

Holmes, W. (1981) 'Cattle', *Biologist* 28(5), 273–9

Janick, J., Schery, R. W., Woods, F. W. and Ruttan, V. W. (1974) *Plant science* (2nd edn), Freeman

King, J. W. B. (1983) 'Pigs', *Biologist* 30(5), 273–276

Lenihan, J. and Fletcher, W. (Gen. Eds) (1975) *Food, agriculture and the environment*, Blackie

Lupton, F. G. H. (1985) 'Wheat', *Biologist* 32(2), 97–105

Moore, P. (1981) 'The varied way plants tap the sun', *New Scientist* 89, 394–7

Pirie, N. W. (1969) *Food resources, conventional and novel*, Penguin

Pullin, R. (1985) 'Tilapias', *Biologist* 32(2) 84–88

Rappaport, R. A. (1971) 'The flow of energy in an agricultural society', *Scientific American* 225(3)

Scientific American (1976) 235(3) (the whole issue deals with food and agriculture)

Sumpter, J. P. and Woods, C. R. C. (1981) 'The trout', *Biologist* 28(4), 219–24

Tudge, C. (1979) *The famine business*, Penguin

The Association of Agriculture produces a series of booklets – 'Modern Agriculture Series'. Send SAE to this charity at 16/20 Strutton Ground, LONDON SW1P 2HP for further details.

Acknowledgments

We should like to thank the following for permission to reproduce the photographs: Farmers Weekly p 47 top; Jersey Cattle Society of the United Kingdom p 47 bottom; Massey Ferguson p 37; Semences Nickerson p 10; Twose of Tiverton Ltd p 40.

Line illustrations by Jake Tebbit.

Index